SAMUEL P. HAYS

A HISTORY OF ENVIRONMENTAL POLITICS SINCE 1945

UNIVERSITY OF PITTSBURGH PRESS

Manufactured in the United States of America

Printed on acid-free paper

10 9 8 7 6 5 4 3 2 1

LIBRARY OF CONGRESS CATALOGING-IN-PUBLICATION DATA

Hays, Samuel P.

A history of environmental politics since 1945 / Samuel P. Hays.

p. cm.

Includes bibliographical references (p.) and index.

ISBN 0-8229-4128-7 (cloth : acid-free paper)

ISBN 0-8229-5747-7 (paper : acid-free paper)

1. Environmental policy—United States—History—20th century.
2. Environmentalism—United States—History—20th century. I. Title.

GE 180. H392 2000

363. 7'05'0973—dc21

For Barbara

CONTENTS

ACKNOWLEDGMENTS

This book calls for a number of "thank-yous" to those who have helped to bring it to fruition. Most of these go to Barbara Darrow Hays, who participated in the venture at every step of the way. There was much hashing and rehashing of ideas, organization, primary points and secondary observations, examples, and how much detail to include. Such matters took up almost daily conversations. But she also contributed much to the writing, helping to unwind cumbersome sentences and especially correcting my ecological misinformation. Her reading over one summer put the finishing touches on the final version.

I do not have the usual thanks to libraries since the book is based entirely on our own archival environmental collection, almost all of which by this time is deposited in the Environmental Archives of the Archives of Industrial Society at the University of Pittsburgh. But I do have to add to this an acknowledgment beyond our first synthesis of environmental affairs, *Beauty, Health, and Permanence,* published in 1987, a "thanks" that goes to the Internet. The range of sources on which our environmental writings are based has been extended mightily by sources on the world wide web, ranging from government documents, both federal and state; to publications of organizations, an enormous number of which maintain web sites; to the readily available newspaper stories, which have enabled us to keep in touch with environmental affairs from coast to coast. Since I am convinced that the electronic world is fleeting and that for future historians much of this evidence will be long lost, I am grateful that it is available to enrich this stage of our environmental writing.

Finally, I am grateful to the University of Pittsburgh Press for being so ready, willing, and able to publish several of my environmental books. Cynthia Miller, the director, who came onto the scene after I retired in 1991, has been more than enthusiastic about what is now my third such book with the press, and Niels Aaboe, the editorial director, has shepherded the manuscript through in most agreeable fashion. I also appreciate the reports of three external readers, all of whom greatly improved the manuscript with their comments.

December 1999

A HISTORY OF ENVIRONMENTAL POLITICS SINCE 1945

1 THE SETTING OF ENVIRONMENTAL POLITICS

Environmental politics has become a major feature of the nation's political landscape during the past several decades. It has involved personal values, local public affairs, and state and national politics and has evoked a significant response from governments and private corporations. As a result, such issues have assumed central importance in the thinking of many Americans: the attempt to preserve nature in the city, the countryside, and the wildlands; the need to control and prevent pollution; and concerns about restraining population and consumption pressures on the finite resources of land, water, and air. These issues in turn have given rise to a host of policies, regulatory programs, and private actions as well as to a complex of subjects, each with its own detail, often its own science, its administrative complexities, and its political strategies.

The result is a welter of environmental circumstances whose intricacies are difficult to comprehend even for seasoned environmental observers. The mass of detail, moreover, provides opportunities for those with anti-environmental agendas to confuse and obscure legitimate environmental concerns in order to counter public support for environmental values. Often those strategies impede an accurate understanding about the roots and context of environmental affairs and their historical development.

Where do environmental issues fit in the current world of American public life? How can one wend one's way through the morass of detail to the underlying environmental realities? How can one develop a clearer understanding of the environmental scene so as to confront environmental circumstances directly and facilitate environmental engagement?

Environmental affairs are, in fact, relatively simple, and one can go directly to their more basic elements. Many writers do this by imposing their own "big picture" on their readers, which leads them into a world of abstracted ideas somewhat divorced from human circumstance. My own preference is to find simpler notions not in the larger philosophies but in the innumerable patterns of concrete human experience and behavior. Who is involved in environmental affairs, what is their personal experience, and what do they seek to achieve? How does one fathom the indifference and routine opposition to environmental objectives, not just on the part of the "interest groups" but also by the nation's leaders? Such are some of the questions dealt with in this introduction to environmental politics.

I start from the experience of people who have something to say about the quality of the land, air, and water around them. They seek out environments that to them are of a higher quality, and they seek to improve the environments where they live, work, and play that they feel have been degraded by development, by pollution, or by other human impact. To understand that experience we must watch two things. One is the state of the environment itself, how it has become degraded over the years because of the increasing effects of human activities; the other is the way in which people have sought to retard this degradation and to bring about improvement in environmental quality.

Both of these factors have played a larger role in our lives over the decades since 1945. We have come to focus more sharply on environmental change as something of importance in private and public affairs. And we have found ways to bring about environmental improvement. We formulate environmental goals and values and implement those goals through science, technology, and both private and public action.

Environmental politics involves three competing forces: individuals and groups motivated to protect and improve the environment, the environmental opposition, and the institutions of policy development and implementation. Here we will outline each of these persistent forces, their origins, direction of thought and action, and role over the years. These forces have constituted the main elements in environmental politics from World War II to the present day. There is little doubt that they will proceed in relatively similar patterns for many decades to come.

The same continuity is displayed in the focal points around which these three forces array themselves: the objectives of public policy; the translation of legislation into regulation; the day-to-day implementation of regulatory rules; the role of science, technology, and economics; competition for the use of the instruments of government—executive, legislative, administrative, and judicial—in decision making; the relative roles of local, state, and federal governmental action.

What began in the early decades of the environmental era after World War II as preliminary initiatives in public policy and decision making gradually evolved into highly developed institutions and political alternatives. If this environmental system has achieved any significant maturity over the intervening decades, it is that it has become more extensive and complex. Here we explore these basic dimensions of environmental politics as the framework through which the vast array of environmental facts, debates, and issues swirling about us can be made sensible.

The customary approach by those who analyze environmental policy, it should be emphasized at the outset, is quite different from what we take up here. Usually environmental issues are thought of as a series of problems to be solved; problems are identified, then possible solutions are outlined and evaluated. This one could call a "problem-policy" approach. The approach used here goes one step further into the context of values and institutions and observes problem definers and solvers as part of the complex of contending political forces. The selection of problems to be emphasized is itself a focal point of political struggle, as is the choice of appropriate research areas and the assessment of results of environmental activity. These questions are the subject of continuing and intense political dispute.

While one can become conversant with the problems and solutions in each environmental field, there is a pattern to the way different values and objectives enter into each arena and by which the various elements of public environmental affairs actually come together and are closely integrated by the three major directions in values and objectives inherent in them: calls for environmental improvement, opposition to those initiatives, and strategies for implementation. The problem-policy approach tends to fracture environmental understanding, but a focus on competing political forces brings the diverse elements together into a simple three-part whole.

For most people, knowledge about environmental affairs is shaped by the media, which have a strong bias toward sensational events and personal drama. But if one is to understand public affairs adequately, one must look beyond the sensational and the personal to the underlying forces that reflect personal and social values and institutional setting. The social, economic, scientific, and professional context is more far-reaching than the single event, and the event cannot be understood outside of that context. At the same time, events can be understood better in historical terms; that is, what happens at one period of time grows out of what happened earlier.

The approach taken here, therefore, is to emphasize context rather than follow events alone. In this way one can understand the setting that sustains the larger relevant political forces, the setting in which much that happens today is both closely rooted in what happened yesterday and, in turn, will shape most of what happens tomorrow. We stress the continuities that reflect incremental development rather than successive dramatic episodes.

A major element of that context is our continual focus on change over time. A single event is one thing, but it is closely tied to previous developments. Hence a historical approach that can reconstruct how things remain similar or differ from one period of time to the next is essential in order to understand environmental politics. Innovation in values, science, technology, governmental policies, and human understanding is slow and evolutionary, rather than rapid and revolutionary. Others may emphasize Rachel Carson, Earth Day 1970, or the nuclear accidents at Three Mile Island and Chernobyl, but this account of environmental politics emphasizes patterns of persistence and change over longer periods of time: how do environmental affairs differ in the year 2000 from those in 1945?

2 A HISTORY OF ENVIRONMENTAL TRANSFORMATION

Increasing awareness of the wider environment in which Americans live has led to much interest on the part of both citizens and scientists in describing that environment. A new breed of historians has arisen to chart and describe such matters, and within the world of science a greater number of specialists have been enticed to examine systematically how the environmental world works. New fields have emerged from botany and zoology, now transformed into varied ecological specializations to explore the interactions of living organisms. Using biology, geology, and chemistry, biogeochemical cycles have been reexamined to determine how human influence modifies these natural cycles.

A crucial task of this desire for environmental understanding is to chart the path of environmental transformation: what is the state of our environment today and how has it changed over the years? The beginning point is the impact of human activity on our finite surroundings of land, air, and water. What has that impact been in the past; how has it increased or diminished over the years; what are the kinds and pace of human influence from one period of time to another as population totals and levels of consumption have grown and technologies have changed?

For many years these environmental transformations went on without much widespread notice or without giving rise to scientific study or public concern. But by the mid-twentieth century, an awareness of the changes emerged to shape a general public consciousness of environmental conditions and to direct an increasing amount of scientific energy into understanding them. Thus were born the two main facets of contemporary environmental affairs: scientific inquiry and public action. To chart all this change requires that we separate these two aspects: environmental transformation as it took place over long periods of time, with relatively little notice or concern, and the rise of public environmental consciousness.

It matters that this increased interest in environmental affairs occurred at a particular time in particular places rather than at other times and places, and among particular people rather than others. These differences help to identify the origins of environmental awareness. People in the past were not completely indifferent to environmental circumstances; protests to environmental conditions occurred, especially in cities. As time went on, these limited reactions evolved into a more widespread interest and gave rise to extensive scientific study and public action. I am interested, therefore, in exploring the precise timing and social roots of environmental awareness as well as the pace of actual environmental changes.

In this chapter I outline the major transformations out of which environmental interest sprang, but do so in only a limited fashion. Here I am concerned primarily with more recent environmental developments rather than a full-fledged history of environmental change; hence that topic will be dealt with only as a background for the larger theme of the book. I divide the historical stages into three parts: (1) the years prior to industrialization and urbanization, in which the human impact on the environment was relatively limited; (2) the first hundred years of urbanization-industrialization in the United States (1850–1950), in which new directions of environmental change can be identified; and (3) the years since 1950, when that change accelerated markedly.

PREINDUSTRIAL SOCIETY

Native Americans and Farmers

Historians have written extensively about the environmental culture of American Indians but less about their environmental surroundings and practices. Within the limited environments that their small populations inhabited, the Native Americans engaged in practices little different from those of the Europeans who displaced them. Their major use of the environment was for food and shelter. Some practiced agriculture in cultivated fields as well as hunted and fished and gathered edible plants. In many cases, their practices exhausted resources and prompted movement to new places to exploit new resources. They used fire to generate browse for deer and to enhance deer populations for hunting, thereby continually interrupting forest ecological processes. In the Southwest they practiced irrigation. These practices reflect not a people "in harmony with nature" but a native people who used their immediate environment intensively. Their comparatively small populations, lack of firearms, and ability to move to new unoccupied areas limited their impact and allowed for environmental recovery. European peoples who displaced them exercised far greater pressure on the environment because of their greater numbers and more powerful technologies.

A second stage in environmental history has been called "agro-ecology," and it represents the beginnings of more intensive agriculture and extraction of raw materials. This analysis has focused primarily on agricultural changes in Europe rather than America, but with the understanding that these changes were brought by

European settlers to America to constitute a new phase of American environmental history—one marked by agricultural settlement rather than more mobile aboriginal occupation. Increasingly intensive agriculture is one of the more significant facets of human pressure on a finite environment. Increasingly intensive use of labor and capital (fossil fuels, pesticides, irrigation, and fertilizers) overcame environmental limits and increased the output of farms and orchards. These intensive practices were used by all sectors of society, from individual farmers to large-scale private enterprise. Their impact on the environment was accepted—with few concerns—by people of their time.

These two earlier stages of environmental history involved far more direct relationships between humans and their natural environment than we experience today. Over time, those relationships became less direct, more impersonal, and less easily perceived. In the later years of the twentieth century, human values and perceptions also changed as the human impact on the environment became more evident and of increasing concern. The earlier acceptance of human environmental impacts makes the emergence of quite different values even more remarkable.

Early Settlement

European settlers in America introduced a distinctive stage in environmental history that emphasized settled agriculture, wildlife hunting, and resource development—all of which increased considerably the intensity of the human load on the environment. Since this occupation was extensive rather than intensive, it proceeded at first without dramatic environmental effects save in limited localities, as was the case with Native American settlement.

The most extensive of these changes was the occupation of lands formerly either sparsely occupied by the Indians or, more commonly, lands that were "empty" forests or prairies. European diseases had generally preceded the new settlers and decimated native populations, a massive environmental change in itself, so that lands once occupied now appeared empty. From the beginning of the nineteenth century, settlement, once largely confined to areas near the Atlantic Coast, rapidly moved westward to carve out farms and establish more concentrated settlements in towns, and in so doing brought about significant change in the environmental landscape. The key watchword motivating such settlement was "land improvement," that is, the process of turning wetlands, forests, and prairies into cultivated cropland. These lands and waters were thought of as wastelands awaiting human occupation to make them produce crops, fuels, or minerals used by settlers as food, fiber, and raw materials. Forests, now in the way, were cut down and the timber disposed of as surplus; scrub land was cleared, prairies were plowed, and swampland was drained so that crops could be grown. Although timber companies were responsible for much early deforestation, land clearing for farming brought about a more comprehensive environmental change.

As deforestation proceeded under the drive to create more farmland, its negative impacts were little noticed. Ohio was an exception. Here the main cause for deforestation was land clearing for food and fiber. By the 1870s over 70 percent of the state's land had undergone "improvement." A few of the state's leaders began to warn of the undesirable effects of such extensive change, and it was here that the modern forest-conservation movement first arose. In later years the role of farming in causing declines in forests and wetlands was almost forgotten as more emphasis was placed on timber harvesting for the market rather than for farmland clearing.

Land clearing and hunting combined to bring about huge changes in the nation's wildlife populations. Early settlers had prized the abundant wildlife found in America. In Europe, wild forests and wildlife were controlled by the royal and noble families, and commoners were denied or given only limited access to them. In contrast, the vast wildlands in America, which were unsupervised by either governments or private owners, were readily accessible to all. Over the years hunting had a devastating effect on wildlife. Some animals were considered dangerous—for example, predators such as wolves—and were not simply hunted but "exterminated." Others such as bear and buffalo were hunted almost to extinction, and by the end of the nineteenth century this was also the case with deer.

Changes in wildlife populations also occurred when wilder habitat was replaced with domesticated habitat around settled areas. Settlement created a more fractured landscape in which large intact forest areas were now divided into parcels by roads, fields, homesteads, villages, and towns. Animals that required large forested areas (such as bears) declined in number, and those that could live, or even thrive, in proximity to humans and human settlements (for example, raccoons) increased. Modern ecological science emphasizes the impact of habitat fragmentation on many species. The first stages, however, came with early land clearing and the substitution of farmsteads for wildlands.

A significant feature of the transformation from wild to domesticated habitats was the entry, even invasion, of non-native species of plants and animals. Often the seeds of exotic plants, mixed in with grains, were brought by the settlers as "green immigrants" to the American landscape. Many species (Queen Anne's lace, St. John's wort) were introduced in feed for horses and then spread their seeds to the farms, forests, and roadsides of the nation's interior. Often foreign birds (house sparrows, starlings) and plants were brought by settlers who wanted to establish a familiar piece of "home" in their new setting. Other species were brought by experimental scientists for research purposes and subsequently escaped (gypsy moth), and still others came as stowaways in transportation devices, such as in the holds of ships (zebra mussels) or on imported timber (chestnut blight).

In the course of westward settlement, some areas were bypassed as being relatively unfit for settlement. Often these were hills too steep for farming or for road building. Other areas were unproductive for agriculture, such as the pine barrens of Long Island and New Jersey. Some swamps were too extensive to be drained. Such bypassed areas were "lands that nobody wanted" and hence remained in an undeveloped state for years and decades to come. By the mid-twentieth century, when undeveloped and natural lands were far more prized, these lands, once rejected, presented major opportunities for wildlands protection and management. Many a land parcel preserved from development by either private or public means could be traced to such origins.

Early Manufacturing

Environmental transformation within the countryside was augmented by early stages of industry, involving raw-material extraction and processing, small-scale manufacturing, new sources of power and energy, early forms of steam transport, and the beginnings of waste disposal from manufacturing. We consider this stage of early industrial development separately from later stages in order to distinguish initial environmental influences from later ones.

Two features of these environmental circumstances deserve attention. Despite the growth of industry, the more natural features of the dominant agrarian environment seemed, to most observers, to be able to accommodate change without severe consequences. Most of the environmental impact was local or regional in scope and hence appeared to be relatively unobtrusive in the larger environment. Artists who depicted the early railroads as fitting into a larger agrarian and natural landscape, rather than dominating and overwhelming it, seemed to be making a statement that the old and new were compatible. At the same time, however, the local and regional impacts of raw-material extraction, manufacturing, or transportation were often striking and extensive.

The impact of the new on the old can be traced most readily in the complaints of those who claimed to be harmed by change. These complaints were brought to the courts, which sought to deal with them in traditional common-law fashion. Those who fostered manufacturing and industry were pitted against those who were adversely affected by these changes. Water power was an important case. Early small-scale manufacturing enterprises, such as flour and lumber mills, were located near streams to use the power of falling water; waterwheels were placed in the rivers themselves or in channels into which the flow was diverted and then allowed to return to the streams. As the scale of production increased, such as with textiles in New England, manufacturers wanted a larger and more constant flow, and to obtain it sought to "engineer" the river to regulate stream flow through dams and reservoirs. These changes reduced the water level below the dams and flooded upstream lands, both markedly changing the aquatic environment of the streams and provoking considerable protest from those whose land was flooded or who lost water and wished to keep the river "natural." Those who wanted to create more intensively engineered rivers generally won out because courts argued that the change was more beneficial to society as a whole.

Equally significant changes came with the industrial use of wood for fuel, but in this case some environmental effects led to conflicts and others did not. In their early years, locomotives fueled by wood emitted sparks that set fire to farmers' fields. In sorting out the resulting protests, the courts at times found that the farmers were at fault because they left combustible material in their fields. The widespread use of wood to make charcoal for the early iron industry, however, created little opposition. Charcoal making required a large amount of wood and usually drew upon younger trees that were smaller and were called "roundwood." Each iron furnace soon outran the geographical range of its supply, motivating the iron companies to purchase large tracts of forest land. At the same time they cut trees when quite young, which in turn grew back to produce more supplies. Though later forest ecologists would observe the significant impact of such practices on forest conditions, rural landowners at the time thought little of it, as it provided a market for their wood.

THE INDUSTRIAL YEARS, 1850–1950

Throughout the nineteenth and twentieth centuries, the three major human pressures on the environment—rising levels of population, consumption, and industrial production—grew persistently and steadily. Yet there were breakpoints in which more than ordinary spurts took place in the acceleration of one factor or another. For the influence of industrial production, 1850 is one such specific point, but the decades of the 1840s and 1850s were a transition period. While population and consumption continued their steady growth during the century from 1850 to 1950, they also took a distinct spurt in the mid-twentieth century, with 1950 as another specific breakpoint and the decades of the 1940s and 1950s as a transition period. We look first at the century from 1850 to 1950 and then at the years thereafter.

The main themes of environmental history from 1850 to 1950 pertain to the evolution of the various human influences on the environment—population, consumption, and industrial production. One might describe the century as one of gradual maturation of the factors that would produce the environmental pressures that increased in marked intensity after 1945 and led to their becoming the subject of inquiry and action in the last half of the twentieth century.

The general directions of industrial growth, and especially its environmental effects, were set prior to 1850, but the two decades surrounding that year marked a rapid acceleration of both growth and impact. One factor responsible for this more rapid pace was the expanding use of coal to replace wood as a fuel in manufacturing and transportation. Coal burning produced air particulates and left deposits, such as lead, in areas downwind of the early coal-using industries—in bogs on mountain tops, in lake sediments, or in layers of snow and ice on the Greenland ice cap. Scientists have been able to track the change from wood to coal by examining those residues. The data identifies slow changes over the years and then marked changes in the 1840s as the use of coal rose sharply. Thus we establish 1850 as an arbitrary but meaningful date that separates a new period from an old.

Population

Population data is standard historical evidence of national "growth," but its meaning in environmental terms is not as clear-cut. Historians have been slow to identify and examine the environmental consequences of population growth from impacts that were regional in scope in 1850 to those that became more pervasive and closely interconnected over the years. One can also outline these environmental consequences by contrasting areas with lower population densities with those having greater densities. Within each state there are similar variations; populations concentrated more in some areas and less in others. Environmental history, therefore, requires a study of population growth that relates it not just to broad national trends but to specific environmental circumstances as well. We can better understand this by comparing environmental change in areas of differing population density, such as urban, countryside, and wildlands areas, and by tracking changes in population density over time. In Michigan, for example, there are three distinct population regions, in tiers from south to north, each relatively equal in geographical extent but varying in the percentage of the total state population from 70 percent to 20 percent to 10 percent; the impact of population in each region has been quite different.

Environmental history differed between settings of decreasing and increasing density. Communities based on raw-material extraction such as mining and lumbering collapsed when the resource was exhausted, thereby reducing population loads. In the rural north above the Mason-Dixon line, population reached a density peak in the last third of the nineteenth century and then declined in the face of competition from more productive agricultural areas to the west, a process that took place first in New England, then in the mid-Atlantic states, and finally in the midwestern states. Farms were abandoned and villages declined. The census of 1900 brought home this change to the nation as a whole and led to considerable interest in the loss of what many considered to be a vital segment of the nation's population. Rural populations continued to decline throughout the twentieth century.

A major result of this rural population decline was the creation of many "lands that nobody wanted." Real estate values dropped, farms were abandoned, and rural property taxes went unpaid. Towns and counties now called upon state and federal governments to shore up their economies by acquiring such lands to be managed for timber production, hunting and fishing, and later a host of outdoor recreation activities. Governments acquired lands in the East for national and state forests, parks, and hunting lands, especially in the northern states of New York, Pennsylvania, Michigan, and Minnesota, and in the Appalachian Mountains. After World War II, previously bypassed lands such as wetlands and pine barrens provided opportunities to establish permanent "natural reserves" amid growing urbanization.

Consumption

Human consumption levels grew steadily over the nineteenth and twentieth centuries to impose an increasing load per person on the environment, until by the last quarter of the twentieth century it was often remarked that the high level of consumption in the United States constituted the heaviest such consumption load of any nation in the world. Through those years consumption went through several distinct stages, from necessities (food and housing) to conveniences (household appliances and cars) and then to amenities (recreation, knowledge, and leisure activities). In the century from 1850 to 1950 the main change was from necessities to conveniences, as electricity and the gasoline motor brought a new dimension to the lives of Americans. The impact of these innovations came first in the 1920s, expanded even during the depression years of the 1930s, and accelerated markedly after World War II.

A significant feature of these changes was the way in which food, clothing, and shelter came to be designed not as necessities but as conveniences and amenities, and in this form constituted the greatest change in consumer-based environmental loads. Increasingly food was advertised in terms of "convenience" as it was precooked and packaged for ease of preparation and consumption; clothing was modified in ways that had little to do with basic protection and more to do with style and consumer preference; homeowners came to expect rooms for each child and several family-use rooms, including dining, living, and recreation rooms. The size of yards and gardens grew. Earlier forms of necessities had produced a significant load on the environment, but these newer forms, with large components of both convenience and aesthetics, increased that load dramatically.

Each new stage of consumption involved a significant expansion in raw-material extraction, and these materials were increasingly gathered from more diverse and distant sources. More intensive methods of refining, greater use of transportation and communication in bringing factors of production and sale together, and far higher expenditures of energy were all required. Each stage of this process of production and consumption was associated with greater, more elaborate, and increasingly far-flung environmental consequences, making the task of tracking the connections between consumption and its consequences even harder. At one time those consequences were near at hand and clearly visible, but they evolved into consequences that were farther away and more difficult to observe. In later years, more serious efforts to track the environmental consequences of consumption came to be known as "life-cycle costing."

Developing a history of the environmental impacts resulting from increasing levels of consumption is thwarted by a tendency among historians to write simply about the manipulation of consumer choices by those who promoted sales. Enticements to greater consumption were, of course, ever present, but they worked their way on human desires that were more than receptive to the overtures of marketing specialists. People wanted higher levels of material goods, conveniences, and amenities. The role of human values in consumption trends in the nineteenth and early twentieth centuries contrasts with later years, when an increasing focus on quality of life sharpened tensions between material and qualitative aspects of human well-being within individuals, communities, and the society as a whole.

Industrial Production

As the number of factories and their size and scale of production increased, so also did their impact on the environment. These effects were known to economists as "externalities." Many communities that had accepted a factory in their midst when it was small found that the same factory, now grown large, produced undesirable waste, smells, and noise. As time went on, factories came to be located not just close to raw materials in rural areas but close to markets and managerial resources, giving rise to urban factory districts. These, in turn, attracted the construction of nearby worker housing and enticed workers to accept less than desirable residential conditions because of the proximity to employment opportunities.

Rapid growth in industrial production gave rise to environmental consequences that often could be traced to the raw materials from which that production was derived. This backward flow of links from entrepreneur to raw-material production is well charted by William Cronon for Chicago, where business leaders in the nineteenth century shaped Chicago's economy by bringing beef, grain, and lumber from far-flung sources to the city and processing them for sale elsewhere. In this way, Chicago entrepreneurs shaped the extensive transformation of the farms and forests in the city's hinterland. Undeveloped land became intensively cultivated and forests were cut down, leading to massive biological changes. Urban demands for food led to massive drainage of wetlands in northwest Ohio, northern Indiana, and southern Michigan. These backward links and their environmental consequences affected a host of communities that were sources of raw material for the new industrial economy.

Two features of the industrial economy played an especially significant role in increasing pressure on the environment: transportation and waste. Innovations in communication evolved in close connection with passenger and freight transport. Mail traveled by horse and carriage and later via railroads; telegraph poles were strung along railroad lines. But these communication devices were of far less environmental consequence than were new modes of transportation, from roads and canals to railroads, and then to highways of ever increasing density and size that evolved at an unrelenting pace. Their occupation of land went on unabated and their use of energy escalated. One could chart, therefore, the successive environmental consequences accompanying the innovations in transportation that came with the growth of the industrial economy as some of the more severe forms of the increasing loads on a finite environment.

Persistent growth in both consumption and production led to the persistent growth in waste. For individuals and families this involved "household waste" ranging from human waste to the ever increasing importance of "consumer waste." For industry it involved the waste produced from raw-material extraction, manufacturing and distribution, and other factory "externalities" objectionable to people. Waste produced by households, tolerated earlier, now came to be intolerable. In rural areas, human and animal waste was disposed of in the open countryside or in outhouses, and discarded household items were dumped in out-of-the-way places and accepted as part of the rural landscape. In towns and cities, however, homes were located much closer together, and the increasing limitation on space led to organized and often highly engineered programs to direct human waste away from where people lived, first via drainage ditches or "dry wells" and then underground sewage drainage systems. In later years, rural people also sought to remove the offending human waste with underground septic systems and to collect household waste in managed "sanitary landfills."

In the more congested areas of cities, waste from processing and manufacturing now also came to be unacceptable. The most offensive of these processing industries were the slaughterhouses, which simply dumped the remains of slaughtered cattle and pigs into streams or burned them near the plant, giving rise to an intense stench. Reaction to the industry was so strong that municipal governments declared that it was a nuisance and must be moved outside the city. Some manufacturing plants also disposed of their waste nearby: iron and steel mills had piles of coal waste, lumber mills produced sawdust waste that was incinerated on-site, and oil refineries dumped their liquid chemical waste onto the ground, allowing it to seep into the groundwater below or drain into nearby rivers and harbors.

As people began to live closer to the offending industries, or closer to each other, waste was "removed" farther from its source. Sewage was discharged into waterways to affect those downstream; factory smoke was directed through stacks up into the air above the surrounding community; and solid waste was taken to an incinerator located beyond residences. In addition to removal, some recycling began to occur, such as the rag pickers who sorted through piles in urban streets to find discarded products that could be sold as raw materials, or those who gathered human waste and carried it to farms near the cities, where it was used as fertilizer. These methods, however, could not keep up with the production of waste. Hence there was a continuous search to find "someplace else" where waste could be "disposed of."

ENVIRONMENTAL TRANSFORMATION AFTER 1950

The environmental tendencies of earlier years, arising from the growth of population, consumption, and industrial production, established a clear historical direction that proceeded throughout the twentieth century. Each of the earlier types of impact became more elaborate and more deeply rooted in the everyday lives and practices of individuals, families, and institutions, and in modes of industrial production.

Population

The rate of population growth within the country rose sharply from 1938 through 1956, then leveled off, and then began another rise in the 1980s. The decline in the fertility ratio, the number of births per mother, was offset by a sharp rise in immigration fostered by more liberal immigration policies. Both legal and illegal immigrants came to the United States, especially from the nearby countries of Central and South America. By the 1990s the nation was experiencing the most rapid annual absolute growth of population in its history. It was encouraged by the desire of employers to secure cheaper labor, by the general attractiveness of individual opportunities in the United States, and by provisions in the law that enabled those already in the United States to bring their families and relatives into the country as legal immigrants. Behind all this was a general public acceptance of an economy and society ever growing in people and productivity.

Consumption

The nation's consumption also grew as a burst of income growth after World War II led to rising levels of consumer spending. Individuals and families could now afford a wide range of consumer products, all of which added to environmental pressures. Most consumption involved considerable energy use, first in the manufacturing process and the transportation of goods to market, and then in consumers' operation of such commodities as cars and appliances. As high rates of population growth combined with a high rate of consumption, the United States displayed an environmental impact probably heavier than any other nation in the world.

The greater distance between consumption and its environmental consequences increasingly depersonalized the links between the two. Earlier the link between one's consumption and its consequences was more clearly visible and often understood through personal experience. Now the links became almost invisible and made it possible for people to expand their consumption without so much as a thought about its environmental effects. Manufacturing and processing usually impacted specific places, but the market that was the starting point for those activities entered into the environmental equation as a broad-based influence in which the place of activity was one's home or place of business, far removed from the place of environmental impact. As long as the problem was evident in the immediate impact of a local factory, it led to public objection and outcry, but in a more global economy much of the environmental impact was more removed from where consumers consumed and hence that impact was more readily ignored.

Industrial Pollution

As the unusable by-products of industrial production accumulated, they added new dimensions to the historical evolution of waste. One was the never-ending search for new places and methods to dispose of waste as tightening policies led to restrictions on previously used places and practices. Waste was often moved from one "sink" to another: water became solid waste as it underwent "treatment" and was spread on land, or solid waste became noxious air and gases through incineration. Polluters increasingly searched for "out-of-the-way" places to dispose of waste, such as under the ground (through injection), the ocean, or new landfill or sludge-disposal sites that might not raise forceful objections. The "search for the ultimate sink" spread waste far from its source to pollute the landscape in new regions and even abroad.

Initially air pollution was thought to be harmful primarily to the communities immediately surrounding the industrial sources, so tall stacks were used to spread it elsewhere. But new knowledge linking upwind sources with downwind effects made it clear that air pollution traveled long distances. Air-pollution episodes that had been only local now became regional as urban smog at times blanketed the northeastern United States and, over time, extended to the South as well. A wide range of chemical pollutants originating from industrial regions of the globe were found around the earth, and by the 1980s chlorinated compounds and other toxic substances had been discovered far beyond their places of origin.

A major new dimension of air pollution arose from the increasing use and dispersion into the atmosphere of synthetic toxic chemicals. These chemicals, not biodegradable, persisted in the air, water, and land; in fact, they were manufactured and used primarily because they did not degrade. But their resistance to biological processes and their persistence made them hazards to biological life generally, humans as well as plants and animals. They concentrated to toxic levels in animal tissue as they moved up the food chain. Because they did not biodegrade, these toxic chemicals migrated widely from their source via water and air and were transported by humans and animals, hence becoming pervasive throughout the globe.

Dimensions of Economic Growth

In the years after World War II, human pressures on the environment became dramatically visible, especially in matters relating to buildings, transportation, and energy. All were closely related aspects of the mounting environmental pressures of the time. They deserve special attention to better understand the environmental transformation in those years.

Rapid construction of homes, commercial establishments, factories, shopping malls, and recreational centers—frequently referred to as "overdevelopment"—was an ever present feature of the prosperous postwar years. Residential areas grew rapidly around cities, and commercial centers grew in connection with them; in many small towns and rural areas, development brought "citylike" influences arising from congestion. The desire for "growth" absorbed the imagination and energies of local and regional leaders, who constantly emphasized the advantages of more population and more jobs. Open areas were continually turned into permanent development, giving rise to many proposals to restrain "growth" or to foster the preservation of still undeveloped land and natural areas. Public discourse displayed a mixture of these two contradictory impulses: public leaders fostered growth and residents participated in it. Now it also seemed that development was being organized and shaped by regional and national efforts. The scale and rate of environmental change had expanded enormously.

In close tandem with increased development came expanded transportation—more and more automobiles, parking lots and garages, airports, and interstate highways. Massive innovations in transportation took place after World War II to accommodate the increasing use of passenger automobiles, trucks, and airplanes, both within and between cities. This expanded consumption was identified from the increase in passenger miles per person, and many environmental consequences resulted: air pollution from automobiles, noise from airplanes, and both pollution and noise from trucks. All became issues, especially where the transportation lines and facilities concentrated in airports, travel routes from home to work in the cities, and trucks on the interstate highways. Conflict arose when land was "taken" by public transportation agencies for highway and airport construction, causing inevitable controversies within the affected communities as to how their land should be used.

Almost every development issue was, in one way or another, an energy issue, since development required energy, gave rise to new modes of transportation that required energy, produced pollution that required energy to mitigate, and generated consumption that required energy both to produce what was consumed and to facilitate consumption itself. An oil shortage in the 1970s gave rise to the related but quite different problem of just how the insatiable energy appetite of the American people was to be satisfied. A decline in economically competitive sources at home had increased U.S. reliance on foreign sources of oil. The ensuing "crisis" generated considerable interest among many Americans in renewable and decentralized forms of energy, such as solar and wind power. Movement toward solar and to a lesser extent wind generation proceeded slowly but surely, yet it was also clear that even these somewhat "environmentally benign" energy sources would not eliminate the huge American energy appetite that fostered increasing consumption.

After World War II the capabilities demonstrated by the atomic bomb were used to develop nuclear power as an alternative fuel in electrical generation by coal, oil, or water. Research and development of nuclear power generation was heavily subsidized by the federal government. Initially touted to be environmentally benign and cheap, nuclear power turned out to be very expensive due to the need for complex safety measures. It also gave rise to the new problem of assuring safety in the required long-term isolation and storage of nuclear wastes. A series of small and large nuclear accidents and spills caused the opposition to nuclear power to grow, so that by the end of the twentieth century no new plants were on order for the United States, though the industry still sold its technology abroad.

Pervasiveness, Scope, and Intensity

The environmental consequences of increasing population, consumption, and industrial production took on many new dimensions in the years after World War II—dimensions of scale, of comprehensiveness, of integration, and of subtlety in their environmental impact that were far less apparent in earlier years. These gave rise to a new era of environmental consequences that reached more deeply into the nation's daily and institutional life, and provided the circumstances within which new environmental values were given impetus and expression.

Especially significant was the increasing reach of environmental effects from the local community to the region, nation, and the entire globe. The depletion of the ozone layer that limits the penetration of genetically harmful radiation into the Earth's atmosphere showed that the adverse consequences of human action extended even to the stratosphere above the Earth.

Equally noteworthy was the way in which the expansion of environmental knowledge and the extension of human environmental experience generated a perspective that "everything is hitched to everything else." As the biology and chemistry of the environment became better understood, the intricate connections between rocks and soil, plants, animals, water, and the atmosphere became more firmly planted in human thinking. These intricate relationships challenged human understanding even as they posed difficult choices for humans amid this complex environmental circumstance.

THE CITY IN ENVIRONMENTAL TRANSFORMATION

As environmental transformation proceeded, the focal point of human loads on the environment was the city. Here the growth of human population concentrated; here also were the higher standards of living with increasing levels of consumption and industry with its environmental externalities. The expanding city was a growth machine calling for ever more development, transportation services, and energy expenditure not only in the city itself but in the region and beyond. Understanding environmental transformation over time, therefore, requires that one outline the environmental stages in the evolution of the city.

Urban environmental historians have customarily considered only the environmental problems internal to the city, such as waste and pollution, and failed to explore the environmental impact of urban centers on the wider region, the nation, and the world at large. All these effects are rooted in the intense population concentration of the city. Here we can only begin to trace the dimensions of this aspect of environmental transformation.

As people congregated in increasing numbers in cities, they created new patterns of human density and experienced the complex tensions between enjoying the benefits of more intensive development and living patterns and witnessing the environmental degradation that accompanies these activities. It is here that the conflict between industry as a source of employment and the human desire for an improved standard of living collided with the desire to live removed from the less desirable areas surrounding factory production. It is here that waste from industry, commerce, and residential consumers came into conflict with the desire for a higher quality of daily life. It is here that the increasing intensity of land use for buildings and streets reduced the open spaces that enhanced urban living. Cities, therefore, define more precisely many of the tensions inherent in environmental circumstance and choice.

From the cities also came the increase in human consumption, with its ever growing impact on the wider environment. Urban consumption increased the demand for raw materials, lumber, coal and oil, minerals, and agricultural products as its own local supplies became exhausted. The impact of urban consumption was also felt in the search for places in the countryside to deposit city waste and in the city residents' use of the countryside as a place for outdoor recreation, vacation residences, and retirement homes. New transportation and communication technologies enabled urban people to penetrate farther and more readily into the countryside, to bring that countryside more fully into the urban orbit, and to facilitate its use and occupancy. The gradual but persistent penetration of urban culture into the countryside is a central theme of environmental history.

Finally, an essential element of environmental history involves the consequences of urban growth for the less settled parts of whole regions, nations, and the world. Those effects have been profound. Demands placed by urbanization on the countryside modified rural land use, altered water cycles, degraded the quality of water in its streams, changed the habitats of plants and animals, and deposited chemicals from the cities onto its lands and waters. Areas of large, intact forest land were carved up into smaller parcels by urban residents who wanted to own their own piece of the woods, with fragmenting effects on wildlife and wild resources generally. Water pollution increased indirectly as urban people consumed more and more mineral resources and stimulated mining-based water pollution, and directly as they deposited waste in rivers and modified the aquatic environment of rivers, lakes, estuaries, and oceans. Cities such as Los Angeles appropriated water resources far beyond their borders. Air pollution originating in cities spread to the wider countryside as well as cities downwind. One of the most dramatic examples was the way in which urban-based automobiles spread lead from gasoline throughout the wider environment.

The urban component of environmental change has been profound and far-reaching, extending far beyond the city. Yet though urban people had a vague sense of the larger environmental connection, that sense truly was vague, and the general environmental consciousness of people and institutions in the cities was quite limited. Hence, though this limited understanding was sufficient to be a major factor in driving a new environmental consciousness, most urban residents remained quite divorced from the impact of their numbers and their actions on the limited environment around them. A focus on the city sharpens our understanding of the nature and extent of environmental change and of both the existence and the limitations of environmental consciousness.

3 THE ENVIRONMENTAL IMPULSE

SOURCES OF THE ENVIRONMENTAL IMPULSE

The increasing public interest in environmental affairs has given rise to debates over the origins of and support for the environmental "movement." Environmental advocates view it as a natural reaction to environmental degradation, an explanation that has some merit to it, but it does not tell us why environmental interests emerged so distinctively in the years just after World War II, when in earlier years people had remained rather passive in the face of environmental change. Opponents of environmental initiatives, on the other hand, seek to argue that environmental advocates come from the "elite" of American society, that they are not an integral part of the social order but the products of cynical manipulators of popular opinion or are people who cannot think wisely and have succumbed to simple and erratic notions about the world. Such explanations fall short of the mark.

The environmental drive in modern society stems from new human values about what people want in their lives. For many decades, even centuries, the desire for more material goods shaped what people thought of as a higher standard of living. In the twentieth century, however, one increasingly heard the phrase "quality of life," a term that reflected a new dimension to "standard of living." The "good life" now referred not only to material goods but also to the quality of the environment where people lived, worked, and played.

This expanded definition of the good life reflected a change that developed incrementally, as one generation of children succeeded an older generation of adults. Changes in human values rarely arise from changes in the lives of just one generation, but typically come as a new generation views life differently from an older one and develops different expectations for their adult future than their parents had. Today's younger generations have developed new ideas about material goods; they desire houses that have separate rooms for each child, for example, or separate living and dining rooms, kitchens, and recreation rooms. These are higher "standards of housing" that often involve quality-of-life objectives as well as simply more space. And with younger generations comes a desire for more travel, recreation, and leisure—opportunities for enhancing quality of life not available to previous generations.

So also, as a part of these value changes, the environment has taken on a new meaning for many Americans. It has long been noted that environmental values are associated with higher levels of education, as are ideas about human health, smoking, and diet. The desire to be free from potentially harmful environmental pollutants is one of these values. Appreciation for the aesthetic values in the natural world and the excitement of learning about that world amid an increasingly urbanized society is another.

The existence of these new quality-of-life values does not mean that they have replaced the more traditional material values or that they ever will. The human world is one in which values compete with each other. In the lives of individuals and families, scarce resources of time, energy, and money call forth continual choices and trade-offs. Some people give more emphasis to material goods and others to quality-of-life experiences. Thus, although environmental values have invaded all levels of the human world and have become influential, they have not become dominant.

As environmental values gained momentum and struggled for more recognition, they have encountered resistance. It is more than misleading to argue that those who shape the institutions of community, state, region, nation, and the world have become fully converted to the wisdom of environmental values. Such values certainly have entered into their thinking, and even what they profess to desire, but in the institutional context of public affairs, they have yet to become as influential as developmental and material values. In this context it seems sensible to recognize the intense institutional resistance to public environmental objectives and the strength of the environmental political opposition.

The evolution of environmental values among the general public provides a continuing progressive direction to environmental affairs. People affirm a desire to have things "better," to carve out continuing environmental gains. Their thinking is future oriented rather than focused on the past, a context frequently conveyed by the phrases "environmental futures" or "vision" statements. Anti-environmentalists seek to define the environmental objectives as retrogressive, as directed toward restoring a romanticized past. They argue that environmentalists wish to "turn back the clock," to live without electricity or in other ways to reject "modern society." On the contrary, the direction of environmental objectives is progressive, based on attempts to create a better life.

Environmental values are an integral part of an urbanized society, and they have grown as urbanization has grown. They are an effort to define an urban-based quality of life: what kind of environment do the people of American urban society want? A host of assessments that rank cities as "places to live" have been published in recent years. Each one uses a different set of criteria to evaluate cities, but in almost every case environmental quality is an important consideration. Annual reports by *Money Magazine* find that respondents rank clean air and clean water as the top two attributes of cities in which they wish to live. Respondents expressed the same preferences when asked where urban people would like to live; the vast majority would prefer a less crowded and more countrylike home environment. Although they are not able to realize their wishes because of where desirable jobs are located, the desire to enhance their environment is strong among urban dwellers.

Those in rural areas who associate themselves with environmental objectives have usually derived their values from the culture of an urbanized society. They can live in less degraded environments largely because the amenities of modern urban-based life—such as electricity, the automobile, the telephone, television, and now the Internet—are available to them. Some have grown up or worked in urban areas or have been exposed through education and the mass media to urban values. From those experiences they have acquired the widespread urban-based desire for a higher quality environment, and this desire often outweighs job uncertainties and prompts them to live in the countryside or in smaller towns and cities.

Sociologists are accustomed to speak of these changes in values as reflecting a new philosophy or "paradigm." There is some truth to this, but one should be cautious about such an argument because it seems to explain people's environmental impulses in abstract terms. In fact the environmental impulse is quite concrete and is associated with the direct engagement of people with their physical setting. At root, environmental affairs in modern life are a combination of changes in values and environmental circumstances; concrete experiences with pollution, natural areas, or crowding shape those values into concrete actions. Environmental philosophies may come later, but they are not the beginning. More people become engaged with political action by experiencing wilderness areas of outstanding natural beauty than by reading Henry David Thoreau and John Muir—although these may come later as people search for wider meaning and explanations.

Observers of this evolution in environmental values have found that the causes for such changes are difficult to establish, yet some conclusions can be drawn. One is that environmental values rise with levels of education and are negatively related to age; they are weaker among older people and stronger among the younger. This observation tells us little about the roots of change but only that the rise in environmental values came about in much the same way as many new values in the twentieth century. Another conclusion is that environmental values are not distinctively associated with income levels. The notion that environmental values are "upper class" has little support in the evidence. They are associated with almost every income level, from the well-to-do who donate land for permanent conservation to the poor who object to their communities serving as sites for pollution-emitting industries and toxic-waste disposal.

If one examines environmental action throughout the nation, one concludes that environmental values are widespread but often latent until some event arises that prompts people who hold these latent values to action. A recurring story with many toxic-waste issues is one in which a mother living a rather traditional, middle-class life suddenly finds in the neighborhood a threat to her health and the health of her children. Such experiences have turned many a politically passive person into an active one, dramatically prompting them to modify roles that were predominantly within the home to new roles of community leadership. When questioned about their lives, their frequent response is, "I never thought I was an environmentalist, but . . ." Hence one concludes that a major feature of environmental affairs is a change in values that takes place in the lives and emotions of people but does not become manifest until a circumstance arises that activates those latent values.

Some insight into the roots of environmental values can be obtained from examining the way in which different geographical areas express different degrees of environmental activity. This can be done by following state and regional responses to specific environmental issues and comparing such matters as their budgetary support for environmental programs, growth in the size and support of state park systems, and progress in replacing older commodity values with new ecological values in forest management. But the most precise regional variations can be developed by examining the voting records in the U.S. Congress over the years since 1970 and ranking those votes by congressional district.

The areas showing the strongest environmental values are New England, New York, and New Jersey in the mid-Atlantic states, Florida in the South, the upper Great Lakes states of Michigan, Wisconsin, and Minnesota, and the three Pacific Coast states. The lowest-ranking states are the western Gulf states of Texas and Louisiana, the Plains states from the Dakotas south through Texas, and the Rocky Mountain states. If one then examines these states in terms of other regional environmental activities, one finds additional variations: in state policies, in the amount of environmental education offered in colleges and universities, in the strength of citizen environmental organizations, and in the degree of press and media coverage of environmental affairs. These reflect regional environmental cultures that pervade broad areas in quite different degrees.

These regional differences can be described in terms of variations in the underlying economies. Those regional economies that are sustained primarily by older extractive industries such as agriculture, grazing, mining, and forestry reflect lingering rural and extractive values and tend to demonstrate low levels of support for environmental activity. Those regional economies that have developed information- and service-based economies that require higher levels of educational skills and are not directly associated with extraction and processing of raw materials tend to have higher levels of environmental support. A crucial feature of these distinctions often has to do with the degree to which economic advisors in these regions advocate the need for high levels of environmental quality and high levels of education in order to sustain their economic future. Areas with high environmental quality emphasize their region as a place to live and enjoy the added benefit of attracting affluent retirees, who bring with them income from retirement funds and investments. In such regions there may be a high level of tension between "old economies" and "new economies."

AESTHETIC VALUES: NATURAL BEAUTY

The first and most lasting environmental value to develop involves an aesthetic and intellectual interest in the natural world. Beginning with the outdoor recreation movement of the 1920s, appreciation for the natural beauty of forested lands, high mountain peaks, and more natural lakes and rivers was gradually shaped into a political movement to protect and preserve such areas from development. About the same time interest in open space and the natural world began to grow in suburban areas and the rural countryside. The focus of such interest increasingly was wetlands, barrens, streams, and valleys that were smaller in scope than the nation's wildlands, but still contributed to creating an aesthetic interest in nature.

One of the first public-policy arenas to display environmental values was the movement to preserve the more spectacular natural areas by establishing national parks, wilderness areas, and wild and scenic rivers. Following the creation of the first national parks in the late nineteenth and early twentieth centuries, the Wilderness Act of 1964, which established a national wilderness system on federal lands, was the culmination of three decades of activity by the Wilderness Society. The Wild and Scenic Rivers Act of 1968, under which some of the nation's larger rivers were protected from development, though equally important, had a shorter history. With the passage of both laws, public interest in nature protection grew, so that candidates for the national wilderness system were extended from the West to the East, involved areas that had been restored as forests, and came to include areas that were much smaller than original proposals. Wild and scenic rivers programs arose at the state level to protect rivers that were smaller and confined within one state's boundaries; the larger, interstate rivers were protected under the national system. As time went on, candidates for both systems were proposed by interested citizens as quality of environment came to be defined in regional, state, even local terms.

Regional interest in the protection of nature went far beyond these larger and more spectacular initiatives. In the nation's suburban areas the task of preserving open space amid rapid urban development came to be defined as an important environmental issue in the late 1950s. In 1961 the federal government provided matching funds to urban areas to purchase land for permanent preservation. A new type of local organization, the "conservation commission," was developed in New England and spread to a few other states; its main task was to undertake surveys of available undeveloped land and to find ways of insulating it from the development pressures of the private market. Later, an even more popular instrument of action was the privately organized conservancy or land trust that acquired lands for permanent protection. Although some conservancies had a longer history, the method became increasingly popular in the late 1970s, and by the mid-1990s there were over eleven hundred such organizations throughout the nation. At first most popular in the Northeast, they also came to be formed in the Midwest and the Far West to counter the impact of rapid development.

The search for natural space to protect led to a focus on areas of special ecological significance, such as wetlands and pine barrens. The wetlands movement developed in the 1960s and grew slowly until by the 1980s both federal and state wetland programs had come to be a central feature of land protection. Governments from the town and township level, as well as states and the federal government, came together to implement the public's desire to protect wetlands. Pine barrens, especially in New Jersey and Long Island, were a second focus of environmental interest, and, like wetlands, were subject to a combination of private action, public land acquisition, and the regulation of development to preserve open space. In some cases, protection of aboveground resources was intimately connected with the protection of underground water supplies.

Much of this initial phase of interest in natural values was associated with outdoor recreation. It grew out of the interest in camping and hiking that arose in the 1920s, accelerated in the 1930s, and took off to an even greater extent in the years after World War II. Wilderness, natural rivers, and hiking trail programs such as the Appalachian Trail were part of this development. But beginning in the 1960s and extending over the next decades, a new interest arose in the biological life of forests and undeveloped lands that went beyond the aesthetics of large landscapes to include the plants and animals within them. This new interest was reflected both in a fascination with ecological science and in the popularity of televised nature programs and photographic images of nature, as well as in nature appreciation generally. Wildlife viewing and photography came to be an important part of tourism, and guides were prepared to direct tourists to the best sites for wildlife viewing.

Public interest in nature led to increased awareness of the specific issue of the protection of endangered species and the more general issue of wildlife protection, as well as the importance of "biological diversity" (or "biodiversity" for short). Programs arose in many states to protect "natural areas" or "rare and endangered species and species of special concern"; these were financed by funds from income tax rebates or the sale of special automobile license plates. In the same way that programs benefiting hunters had long been financed through hunting license fees, now those interested in "nongame" wildlife financed the programs in which they had a special interest. A host of measures were adopted at the state and local level to finance the acquisition of lands for public ownership and protection, and much of their popularity came from the new interest in the ecology of areas that were more natural and less developed. The message conveyed in these ventures was to keep it that way.

This widespread interest in the natural values of wild and undeveloped land was a new development in American society. In the nineteenth century, wilderness areas had negative connotations as dark and dangerous places; now they were prized for their uniqueness amid an increasingly urbanized society. Earlier, when wetlands had been considered wastelands, the only sensible public policy had been to drain them for farmland, and state drainage laws had contained preambles about their uselessness. But now the same lands came to be thought of as valuable both for their role in flood control and water storage and for their biological uniqueness. At one time wild predators such as wolves and grizzly bears were considered only as threats to be exterminated, but now they came to be prized, and the remaining populations were protected so as to foster their increase.

Some sought to denigrate this increasing interest in the natural world by associating it with a desire to return to some more primitive state of existence that was rural and without amenities. The movement to protect nature was, they argued, simply a yen to reestablish a long-lost past. But the interest in nature was forward-looking, an attempt to improve the quality of life of urban Americans. The aim was not to discard urban amenities but instead to add to them by enhancing the role of nature in an urbanized society. Those active in promoting nature protection envisioned a future in which their children would be able to enjoy a quality of life in which nature played an important role, similar to that which they had experienced. Natural values were closely entwined with the values of a material standard of living, and not in contrast or contradiction to them.

HEALTH

Among the complex of environmental values was a concern for better health and the role of a less polluted environment in fostering that objective. It was not just that the environment had become less healthy as time went on but that ideas about what constituted a healthy life had changed. For many years the main emphasis in health had been to reduce the incidence of death. But as infant mortality declined, as antibiotics reduced the dangers of bacterial infections, and as people lived longer, greater emphasis was placed on improving general health and well-being during one's life. In medical terms, this involved a shift in emphasis from mortality to morbidity. Environmental considerations entered into both aspects of human health. For mortality, the main focus was on cancer-inducing environmental agents; for morbidity, the emphasis was on the way in which environmental agents limited human development or reduced day-to-day wellness.

Some environmental conditions were long understood as threats to human health. In the nineteenth century, physicians often advised their urban patients with lung disorders to move away from the smoke of the city to some area with cleaner air. Sanitariums for tuberculosis patients were located beyond cities, and one of the justifications for establishing public forests in the early twentieth century was to provide cleaner air for rest and recuperation. As the germ theory of disease was accepted in the latter part of the nineteenth century, urban water supplies were treated, which led to a marked decline in water-borne disease, and milk was pasteurized to prevent the spread of milk-borne tuberculosis. For many years bacteria were the major focus of new treatments and of medical science in general. But as these threats to human health were reduced, other conditions of human health came to the fore.

The case of lung cancer was particularly significant in reflecting a change in how health conditions were thought about. In earlier years the major medical focus in lung problems was tuberculosis. Environmental factors, such as exposure to air-borne chemicals and metals, was given only secondary consideration. Mine operators, for example, argued for years that the lung problems of miners were due to tuberculosis rather than conditions in which miners worked. But as tuberculosis came to be treated effectively, other lung problems such as cancer became more visible. Workers exposed to asbestos fibers had long been subject to a lung disease known as asbestosis, but as this condition was treated effectively, lung cancer from asbestos exposure received more attention. So it was with a considerable number of environmental disease agents.

In the middle years of the twentieth century, environmental health was thought of largely as occupational health. Workers were the main group studied for treatment. As the environmental era advanced, however, a larger share of the population received attention. This included women, children, and the elderly, who collectively became the main focus of community health programs. As community health emerged in the 1960s as a new topic for inquiry and action, experts in occupational health were drawn upon for scientific and professional advice. But it was soon found that those who had focused primarily on occupational health and environmental hazards were not prepared to deal with community health issues. Gradually a new area of human health evolved that emphasized the environmental health of children, women, and the elderly.

One of the first areas of concern was children. Within pediatrics there had long been a debate about whether, for medical purposes, children were simply little adults with similar diseases requiring treatment similar to that required for adults, or whether their physiology was different enough to require separate and distinct observation and care. A number of tense and volatile environmental controversies emerged as the focus on children came to the fore. One environmental health problem of particular concern for children was lead. Lead at high levels of exposure was long known to be a hazard to both workers and children, but the effects were thought to be temporary and easily treated. However, some pediatric scientists thought lead had more permanent effects on the neurological development of children, and new knowledge about this began to accumulate in the late 1970s and to gain wide acceptance in the 1980s. It provided considerable support for eliminating lead in gasoline and reducing exposures from drinking water, paint, and other sources. A major result of this long-term issue was to emphasize the uniqueness of the developing child, including the fetus, for whom exposures of a given magnitude to toxic substances had far greater adverse effects than similar exposures in adults.

An even more extensive medical effort was devoted to the elderly. In this case, the issue was increasing life expectancies and the emphasis on a healthy and rewarding life for older people. If one retired at 65 or 70, one could expect ten to fifteen years of fairly active life thereafter, and there was considerable interest in both prolonging life and enhancing its quality. Much of the early interest in air pollution had to do with lung diseases of the elderly—for example, emphysema—that were worsened by air pollution. As time passed, there was increasing recognition of smoking as a cause of cancer later in life. Smoking was also implicated in heart disease and a variety of ailments other than cancer. With regard to the elderly, there was the tendency to argue that diseases were simply the product of natural decline in physical condition, the result of "old age," and hence to de-emphasize any environmental causes. Some even argued that for this reason people over sixty-five should not be included in cancer statistics. But, as with children, increasingly long-term environmental exposures came to be recognized as the cause of some of the health problems of the elderly.

For women, environmental issues were an especially direct and significant concern. On the one hand, the growing incidence of breast cancer, though as mysterious as cancer was generally, seemed to have a direct connection with synthetic organic chemicals that concentrated in breast tissue. On the other hand, varied reproductive disorders increasingly came to be associated with chemical agents. One example was the increased risk of vaginal and cervical cancer that occurred in women whose mothers had been given the drug diethylstilbestrol (DES) during pregnancy. In the early years of these concerns, problems with reproduction were associated largely with female disorders, but increasingly the role of lower sperm counts and sperm viability as well as other disorders gained visibility. The advance of environmental health science over the 1970s and 1980s brought these issues into sharper focus. By the early 1990s a mature science had taken shape that drew on studies in many fields to understand how synthetic organic chemicals disrupted hormones in humans and vertebrate animals generally, affecting not only adults but fetal and child development as well.

The steady progress of scientific knowledge about human health and the implications for it from chemical contaminants was marked by two major directions. One was the shift in focus from the effects of high-level exposure episodes to the effects of persistent low-level exposures. The other was a shift in emphasis from studying the short-term effects of exposures to examining their long-term chronic effects. These shifts gave rise to new perspectives and knowledge about human health problems and greatly expanded the field of environmental health from an initial focus on cancer to a broader look at developmental, hormonal, neurological, and immune-system problems. At the same time, they expanded the subjects at risk from those exposed in the workplace to the entire range of people exposed in the community. As one looked back from the threshold of the twenty-first century at the previous forty years, the change in perspective, only barely observable on a year-to-year basis, was enormous. One could not safely predict the direction scientific explorations thus set in motion would take in the future, but one could already discern that they were an integral part of the new environmental era.

Along with the new ideas and values about health and health science came new health treatments and advice. Sometimes the scientists led the way, as they had in determining the effects of lead on child development. Often, however, the public was ahead of the scientists, as in its belief that some synthetic chemicals were detrimental rather than beneficial to humans, a view that was greatly resisted by the chemical industry and the scientists who worked for it. One could understand this change, just as one could understand the interest in a more natural environment, as another example of the long-term drive by people for a "better" life. There was now a view that human conditions and experiences other than material goods were important to the "good life," and a health-promoting environment was an integral part of that goal. Such an objective was not firmly defined but was part of a continuously evolving idea in America about what was "better," an idea redefined from time to time, expanded and elaborated, but one with a fairly clear sense of direction, a continuous striving for improved quality of life.

PERMANENCE: ECOLOGICAL STABILITY

Quality-of-life objectives and values took on a more visible and problematic focus as people sought to restrain the intensity of the development forces around them on the grounds that they brought too much instability to their lives and their environment. Economic forces have been agents of radical and fundamental change throughout American history and have led to the disruption of many patterns of living that were formerly well established. Resistance to such changes has been continuous and persistent over the years but rarely successful. In the twentieth century, people witnessed an acceleration of the pace at which natural lands were being transformed into developed lands. The experience was especially acute in the urban-metropolitan areas, where many a family who had decided to move to the outskirts of the city or to the countryside because it was more pleasant soon came to find that massive numbers of others were choosing to do the same thing. Development now destroyed what they had sought to preserve. The conflict between these two forces—those who promoted development with its massive environmental changes, and those who considered the changes degrading to their environment—came to be one of the major tensions of the environmental era. It took varied forms and was the basis of much community thought and action.

The search for stability often took the form of a contrast between the natural forces that supposedly worked in one direction and human intervention that worked in the opposite direction. In the environmental era a widespread view maintained that a mindless human intervention that did not take into account and respect natural forces was a major source of environmental degradation. Some observers tended to sharpen this emphasis on a balance between the natural and the human into antagonistic, black-and-white forces, but most environmental strategies involved rather pragmatic and practical actions to alter the balance of previous years to a more acceptable one. The continuing debate about what was natural and what was human tended to obscure the fact that the resulting ideas and actions were invariably the product of human ideas about what conditions made a desirable human environment.

It was widely felt that a more desirable "balance" required a larger role for nature protection in the urbanized world. Many spoke of this as a "balance of nature," which to them involved not an idea about the way nature worked but a statement about the appropriate balance between nature and humans. This idea often carried with it a fixed notion as to what that natural condition should be, and it was quite easy to think that in some fashion natural forces contained within themselves tendencies toward stability.

A series of environmental mishaps led to an increasing understanding of the way in which human actions could often cause unanticipated and unpredictable environmental changes. Some of the most obvious cases involved building on coastal areas vulnerable to ocean storms, or on floodplains subject to periodic flooding. Many other examples came from the biological world: agricultural clearing destroyed the natural systems of wetlands; ever more intensive agriculture destroyed fencerows that provided habitat for native pollinating insects; and forest clearcutting destroyed plant and animal life that depended on forest habitats. Ecologists spoke of such human intervention as disturbances, compared them with natural disturbances, and tried to understand the differences in recovery from natural versus human disturbances. Now that these changes were considered undesirable, a move arose that more should be known about such sequential natural changes before human intervention was undertaken and that the pace, scale, and form of intervention should be modified.

From such experiences as these arose the relatively new term "environmental protection." For some this led to the idea that "nature" was "all wise," but to most it was simply a reminder to "look before you leap," that is, to find out how the natural world works so as to avoid miscalculations that might backfire. The admonition that came from the new environmental perspective was to "design with nature," as in the title of a major work by landscape architect Ian McHarg.

One of the most salient examples of human intervention causing disturbances in natural processes had to do with the impact of waste disposal on natural cycles. In the natural world, chemicals continually cycled through the environment. A certain balance had developed over the course of time between sources of chemicals and the ability of the environment to reabsorb them. Under the stress of human waste disposal, natural cycles now became overloaded so as to change the air, water, and land where chemicals were deposited. The changes included increasing acidification in lakes, streams, and bays; the impairment of wildlife reproduction and development due to exposure to pesticides; and threats to human health from excessive nitrogen in groundwater. It seemed quite logical that if such overloads in natural systems had caused problems, action should be taken to bring about a reduced "load" that would right the balance.

For many people in urbanized America, the search for more natural surroundings in the metropolitan fringe was continually foiled by the intense development that grew up around them. They were drawn to the city by its job and recreational opportunities, but they were drawn to areas outside the city as places to live. These preferences were shaped by their experience with development activities that were destroying what they had known earlier. Woodlands once enjoyed for outdoor recreation or merely as a scenic part of the community were cut down for new subdivisions and were gone forever. Hunters found that favorite woodland haunts were now off-limits either because those who established permanent or weekend homes in the same areas posted their lands against hunting, or permanent development had destroyed the areas as wildlife habitats. Many families had sought a quieter life with a new home on the fringes of the city only to find that over time the number of houses increased, shopping areas grew, roads improved, and traffic increased. Older residents were filled with nostalgia about what it was like "when we first moved here."

In small towns and the countryside, this instability was underscored by the threats to a long-standing way of life that seemed to come from the wider world—a dam proposed for a river that would flood ancestral homes, graveyards, and farmlands; a waste-disposal site in or near one's community that presented a new health threat to one's family; a new industry that would attract so many vehicles as to generate congestion, noise, and danger. Intensive development was always high on the agenda of local officials and chambers of commerce that spoke for entrepreneurs with a commitment to economic growth. Hence although many small towns and rural areas enjoyed considerable community stability, they continually faced forces that threatened to undermine that stability in the name of progress.

The business community responsible for many of these changes is usually labeled "conservative." But to those who observe long-term historical change, the forces generated by the business community have produced massive and radical change, and those who seek to protect themselves against such change are more appropriately called conservative. Hence it is quite common for those who seek to maintain stability in their own lives and in their communities to think of themselves as defending traditional ways of life against the radical forces bearing down upon them. Among some this leads to a desire to protect their communities against developments thought to be harmful. Middle- and upper-middle-class communities often have the ability to ward off such intrusions simply because their level of income, education, and political influence cause them to be avoided by potential developers. But those communities with more limited resources are placed in the position of having to organize to protect themselves.

Ecological and environmental instability was driven home to most people through direct experiences such as the loss of natural lands in an urbanizing world, the effects of disturbances and pollution on natural processes, or the relentless intrusion of development in communities that sought a quieter and less frenetic daily life. Such experiences gave meaning to the search for stability in a world in which economic forces generated continual instability.

To those who thought in larger terms, these disruptive forces encompassed not just the world of individual and community experience but also the larger world of countries and continents, even the entire globe. Environmental activists continually pointed to limits in the air, land, and water resources of the entire Earth. They argued that populations, economic growth, consumption, and creation of industrial, commercial, and household waste could not go on expanding forever. Continuing growth amid finite limits in land, air, and water generated a continuing instability in individual nations and the entire world. Such views as these led to the conclusion that preserving human society on the Earth required significant reductions in the pressures on the limited environment.

4 NATURE IN AN URBANIZED SOCIETY

One of the most persistent themes in the search for a higher quality of life is the desire for a more natural environment amid the built-up circumstances of modern life. There are three parts to this search. The first is the desire for more natural lands, plants, and animals in one's environment to provide a more natural setting for daily life than the highly developed settings that have increasingly become the norm. The second is the belief that pollution has created an environment that is less attractive and less healthy than one in which more natural biogeochemical processes are less influenced by human activities. The third is the perception that environments that are less crowded with people, cars, and noise enhance the quality of life and are, therefore, desired objectives of private and public policy. Each of these themes enters into the widespread and oft-repeated preference for a more natural environment.

Amid this search for more natural elements is an equally forceful drive to understand how the natural world works and interacts with human influences. To those who participate in it, this search for knowledge has combined the intellectual desire to know with one's mind how the natural world works and an aesthetic appreciation of that world. Hence in today's environmental era a wide spectrum of people participate in the new environmental science, beginning with a desire to learn more about environmental issues in the early years of informal and formal education and continuing throughout life. Activities for citizen participation ranging from monitoring environmental changes to reading books reflect these combined intellectual and appreciative interests.

The search for a more natural living environment also involved a debate over what is meant by "natural." The environmental intelligentsia, represented primarily in the academic, publishing, and writing worlds, often argued that the ideal of a primitive nature lay at the roots of environmental aspirations. This notion of reverting to a primitive kind of nature was belittled by the anti-environmentalists on the grounds that a "pure nature" or "pristine nature" never existed and was merely a figment of the human imagination. Within the environmental intelligentsia itself, some writers took the same view, arguing that by searching for a pristine nature that had never existed, the nature advocates had undermined their own objectives.

The type of nature that was most sought after was far less absolutist and far more proximate. It referred more to an acceptable level of development within one's own experience than to an idealized world of nature defined by some "original" state. The specific definition of this more proximate "state of nature" underlying environmental objectives varies with circumstances: sometimes it refers to small pieces of undeveloped land within an urbanized world; at other times it means a more natural setting in the countryside such as a wetland or woodland that contrasts with more intensively used farmlands; at others it involves a relatively undeveloped wild area that displays a more varied flora and fauna than an intensively managed timberland; or it can refer to naturally occurring geological formation, such as the undeveloped canyon country of southern Utah. Each of these is defined as more natural only in contrast with more developed lands in the respective local area, not because it conforms to an idealized or "original" nature.

The environmental intelligentsia added to their conception of "nature" the notion that it was distinct from and in conflict with modern urban society. Advocates of nature, they argued, rejected the modern world of cities, technology, and material standards of living. Such views are tempting among those who think in terms of well-ordered and logical systems of ideas in which absolute opposites abound. But the values and directions of thought and action among nature advocates reveal that they do not reject the modern world but rather seek to enhance the role of nature within it. Their starting point is not the natural world to which the urbanized world must adjust but a modern, urbanized world that is here to stay and within which they seek a larger role for nature.

WILDERNESS AND THE WILD

In the late nineteenth and early twentieth centuries, the national parks set the tone in the search for ways to incorporate more of nature into an increasingly urbanized lifestyle. State park systems established after World War I, especially in the more urbanized states, grew in popularity during the 1930s, and their use accelerated even further after World War II. These parks were places for relaxation and recreation on the one hand and for the protection of natural phenomena on the other. National and state forest lands became major destinations for outdoor activities, and the administering agencies began to develop additional outdoor recreational programs after World War II. In almost every case, a park's origins can be traced to new ideas and values rooted in urbanized culture and shared by advocates who were seeking not to abandon cities but to reach out to nature for recreation and aesthetic appreciation.

For writers about environmental affairs, the vast forested, undeveloped areas of the West have long dominated ideas about "nature." Their central point of departure has been the "wilderness movement," which had a goal to establish areas in which human influence is relatively unnoticeable. Advocates of the movement succeeded in establishing the National Wilderness Preservation System, as described earlier, in the Wilderness Act of 1964, and thereafter were engaged in adding acreage to that system. The largest addition occurred in Alaska as part of the Alaskan National Lands Act of 1969. In later years a new drive added parts of the dry, rugged lands of the West administered by the Bureau of Land Management.

The wilderness movement, which originated with a small group of easterners of considerable means, played a significant and expanding role in the search for nature. It was this group—whose leader has usually been taken to be the outdoor enthusiast Robert Marshall—that organized the Wilderness Society in 1937 and carried on the two-and-a-half-decade fight for wilderness areas that resulted in the Wilderness Act of 1964. At first this small band of wilderness advocates identified only a few large areas in the West to be part of that new wilderness system. But almost as soon as the act was passed, an expanding group of new wilderness enthusiasts began to demand that other areas be included in the system. Thus began a concerted effort on the part of this new wilderness public that greatly expanded the acreage in the system. In the process, they defined more elaborately and more precisely the values inherent in their drive.

In the late 1960s, westerners, first in Montana and then elsewhere, began to argue that areas not originally identified by the Wilderness Society should also be included in the wilderness system. These came to be called "de facto" ("wilderness in fact") rather than "de jure" ("wilderness in law") areas. As a result of their efforts, a new set of demands for wilderness designations arose. This expanded vision was not readily accepted by those who had led the first stage of the wilderness movement. They felt that the newly proposed areas were not of sufficient wilderness quality to be included in the system. Oregon was a case in point. Here the first wilderness proposals were confined to the Cascades and were an outgrowth of the interests of hiking groups and the Sierra Club. But as time went on, Oregonians interested in both the Coast Range to the west and large forested areas in eastern Oregon began to propose areas there for the National Wilderness System. Amid opposition to this plan from earlier wilderness leaders, champions of the new areas formed their own organization, the Oregon Natural Resources Council, to press their goals. Drawing on supporters from these proposed new areas, they formed an effective political coalition that succeeded in greatly expanding the Oregon component of the national system.

In a similar fashion, wilderness designations were established in the East. Here the U.S. Forest Service argued that only "virgin" areas, that is, those forests never cut by white European settlers, were eligible, thus almost all of the eastern national forests were off-limits because they were "second growth." But to eastern wilderness advocates, the relevant part of the law was the phrase that defined wilderness areas as lands where "human intrusion was relatively unnoticeable." In other words, *how* they came to be was not important, but *how they were now* was. In state after state, eastern wilderness advocates began to gather support for their plans from their elected representatives in Congress. This led in 1974 to the Eastern Wilderness Act, which designated some eastern areas as part of the National Wilderness System and identified others as potential candidates. The U.S. Forest Service never modified its formal opposition to wilderness that was not "virgin" but simply accepted the decision of Congress to add such areas to the wilderness system.

The common thread among all of these designated wilderness areas had less to do with "pristine virgin forests" and more to do with undeveloped lands. The common value at work was an aesthetic one: the attractiveness of natural beauty amid a developed society. Advocates showed little desire to leave the urban society in which they were rooted, but rather wanted to continue to live in more densely settled areas with wilderness areas available nearby. The development of affordable photographic equipment after World War II enabled people to capture these forms of the natural world themselves to share their excitement and experience with others; it also predisposed them to purchase magazines or books with color photography and to watch wildlife television programs. In wilderness activism, the first impulse often was to capture the visual grandeur of an area on film and share it with others, usually with a slide show; that beauty frequently was enough to convince skeptical government administrators and legislators to support preservation of an area.

Those in the forefront of the wilderness movement were driven by their desire to experience outdoor recreation in the wild such as camping, hiking, canoeing, and river rafting. A major component of the wilderness movement was the perception of wilderness areas as undeveloped wildlands that users could enjoy and where they could test their physical skills, but it encompassed a much wider range of wildlands recreation activities. By the late 1950s a National Outdoor Recreation Resources Review Commission was preparing a report that advocated major outdoor programs to facilitate hiking and river recreation. This drive came to fruition at the federal level when Congress passed the National Trails Act and National Wild and Scenic Rivers Act, both in 1968, in which national trails were established on federal lands, and rivers flowing through those lands were designated as federal wild and scenic rivers, protected from development. Many states followed this lead to build trails on their state lands and designate rivers within their borders as protected wild, scenic, or recreational rivers.

THE VARIED SEARCH FOR NATURE

A host of other attempts to bring nature more firmly into modern society drew upon a wider range of supporters and took on a wider range of forms and activities. For the most part these efforts had few intellectual advocates, as they were outside their vision. Moreover, they were not inspired by wilderness ideas or writings, but more by their own values from their own distinctive settings of home, work, and play. They set the tone for "nature advocates" far more than did the wilderness movement. They can be sorted out in terms of the degree of balance between wild and developed lands inherent in their efforts, ranging from the city to the suburbs to the countryside to the more intact wildlands. This continuum also spread across the degrees of human settlement from the more to the less intensely inhabited.

Hunting

The initial and most long-standing thrust to protect and conserve an aspect of the natural world focused on wild game and came primarily from urban-based hunters, who felt that the sport was threatened by overhunting. The nineteenth-century reaction against commercial hunting led to attempts to establish public programs for sport hunters to help restore game populations. One such program involved a plan to propagate game animals "artificially" by stocking areas whose wildlife had been depleted over the years. This initiative on the part of city-based hunters was not always appreciated by those who lived in the gameland areas. Nevertheless, such drives led to massive restorations of deer and turkey, favorite game animals, and later to black bear populations in the forested areas of the East.

Population restoration also involved attempts to establish refuges or game lands. These often carried the connotation of being "public hunting lands" available at low cost to hunters generally, in contrast to the private hunting "preserves" that were acquired and maintained by the wealthy. In the years after World War II, much of the drive to establish public hunting lands came from those who had experienced in their own lifetime the transformation of wildlands where they had previously hunted into developed settlements in which hunting was now opposed and often prohibited. Hunters were far more prone to speak of the detrimental effects of human "overpopulation" than were most groups in society.

Nature Appreciation and Study

Alongside the interest in hunting arose a very different attitude toward nature that involved nature appreciation and study, and by the 1990s it had outdistanced hunting in the number of people who participated. This attitude toward wild animals was known as "nonconsumptive" use or an "appreciative" approach to wildlife, and among its proponents the term "game" often declined in popularity to be replaced by the term "wildlife." Nonconsumptive use took the form of wildlife watching, which in many states gave rise to "watchable wildlife" programs along with guide books as to the best sites for viewing wildlife. The objects of appreciation ranged from whales on the Pacific and Atlantic coasts, to migrating hawks along their flyways, to butterflies in newly established butterfly gardens, to deer, birds, and other animals. Direct watching with binoculars rose in popularity, as did the hobby of nature photography. At the same time, nature programs on television attracted a wide audience, and "nature" and "wild bird" stores grew up to cater to this new market for wildlife appreciation.

This appreciative attitude toward wildlife was reflected in new terms that entered into everyday language, such as "ecology," "biodiversity," and "ecosystem." These terms referred to the variety and complexity of the plants and animals in nature and their web of interactive elements. New links were forged between the scientific disciplines that continued to study and explore ecological systems and the general public that found those biological elements to be fascinating objects to observe and appreciate, to study and understand. What was particularly striking was the way in which many elements of the plant and animal kingdoms that earlier had elicited little or negative interest now took on more positive images under the influence of human curiosity and appreciation. Animals such as snakes and spiders, long feared and disliked, were now studied and respected for their particular contributions to the web of life. Invertebrates generally and insects in particular benefited from this new appreciation for all living things, and museums devoted to them were established.

Nongame Wildlife

Interest rose rapidly in animals that were not hunted, such as bald eagles and hawks, songbirds, loons, butterflies, and invertebrates generally; these were often lumped together in what were called "nongame" wildlife programs. Some interest in this new world of wildlife was expressed at the federal level, but though Congress passed a nongame-program enabling act in 1980, it was never funded. Hence most such programs evolved at the state level, where varied new funding sources such as lotteries, income tax refunds, or special wildlife license plates generated funds to protect a wide range of wildlife. In most states, existing forest and game management agencies were cool to natural-area programs, so separate agencies often arose to foster them. As the public support for hunting seemed to decline, game agencies were reluctantly willing to accept nongame wildlife programs, but rarely were they able to go beyond accepting new monies for these programs to the more painful task of shifting resources from the old to the new. Thus state and federal appreciative wildlife programs lagged far behind the public interest in them.

Endangered Species

In the 1960s the growing interest in more varied species of plants and animals led to a special focus on their actual and potential disappearance. The decline and disappearance of species, the most well known of which was the passenger pigeon, came to preoccupy both professional biologists and the public. In the first federal endangered species law, passed in 1964, larger vertebrates were the main subject of attention, but within a few years the interest of both scientists and the public began to expand to include the lesser animals—even invertebrates—and plants, and these were brought under the protection of the endangered species program in 1974. Federal, state, and private landowners were all required to protect federally designated species. At first the programs focused on the species themselves, but the law stated that habitat protection was essential to species protection, hence the connection between species and habitats was crucial. That idea had been implicit in the shift from "bag limits" to habitat preservation in the programs designed to protect game birds as early as the 1920s; now it became an integral part of the program to protect rare and endangered species as well.

Predators

One of the more remarkable transformations in attitudes toward nature involved predators, those large animals such as grizzly bears, coyotes, mountain lions, and wolves that preyed on other animals. Because their prey was often livestock, earlier predator-control programs had sought to eradicate them completely. Both federal and state programs provided rewards or "bounties" for killing predators. Now, however, the public as well as ecologists came to believe that predators played a significant role in natural systems, providing controls and checks on excessive animal populations. Consequently new programs arose that were designed to restore and sustain predators. Some of the more dramatic cases took place in the Rocky Mountain West, where the large areas required for predator habitat were available. Programs to restore the grizzly bear and the wolf to the northern Rockies were especially noteworthy for the high degree of support they garnered from the urban public of that region and elsewhere. Successful wolf restoration programs in northern Minnesota and in the Carolinas also met with widespread public support.

Specialized Animal Groups

A distinctive form of organization on behalf of nature were the groups that focused on a single species, such as bears, deer, elk, loons, mountain sheep, bald eagles, or wolves. In each of these, a group of people devoted to the welfare of the target species gathered together to support research, education, and appreciation. People paid annual dues to the organizations, and they traveled to research and viewing centers such as the Big Horn Sheep Center in Dubois, Wyoming, or the International Wolf Center in Ely, Minnesota. Through the International Loon Association they monitored loons on northeastern lakes. They joined the Xerces Society to focus on the welfare of invertebrates, and Bat Conservation International to specialize in bats. The long-standing bird counts organized by the Audubon Society were now copied by those who joined in the annual July 4th butterfly counts. Most such organizations had a research focus, usually conducted at a specialized place, such as the Raptor Center in southwestern Idaho, which served as the gathering place for those with an avid interest in raptors. The range of such specialized interests can be charted readily from the relevant books on the nature shelves of book stores.

LANDS FOR NATURE PROTECTION

The attempt to create and sustain more of nature amid the growing urban society required, above all, land on which plants, animals, and whole ecosystems could live and survive. Hence a wide range of endeavors arose to foster nature protection on both private and public lands. These efforts enjoyed considerable popularity, as evidenced by the great number of land conservancies that arose to sustain nature and the frequent attempts to foster public programs to do the same. The latter ranged from simple attempts to preserve open space from development to designating protected areas selected specifically for their ecological characteristics.

Open Space

The initial interest in keeping some elements of nature in the urbanizing environment was advanced by simply maintaining open space: some lands were protected from development and kept as more natural and open areas. This effort was a direct response to development pressures. As cities grew, people continually sought open and undeveloped lands on the urban periphery. Often it was not until some of the last remaining tracts of open land, usually formerly farmland or woodland, were proposed for development that a community energized itself to focus on open space as a valuable present and future asset for the community. By 1960 federal funds were made available to municipalities to acquire open space. Because land was cheaper on the urban periphery than within the city, purchases were often for parks and open areas just beyond the more built-up areas.

The Land Trust

In the 1980s a new form of nature protection received increasing support: the land conservation trust. Land trusts arose first in New England, New York, and New Jersey, but then gradually spread to the Midwest, the South, and the Pacific; as the Mountain states became more urbanized, land trusts developed in that region as well. By 1997 there were over eleven hundred land conservancies, each organized through private efforts and owned and maintained as private entities. Private land trusts were considered to be more effective in protecting land from development than were public agencies, which seemed to be overly influenced by developers. A host of preservation strategies were devised; one of the most popular was the conservation easement, which ensured against development through deed restrictions. Conservancies drew upon many individuals who gave land and money under the general impulse of "saving something for our children," or "preserving this area as we have known it." Their vision was geared to the present and the future, not to some remote past. Rather than exposing their land to the vagaries of the short-term financial markets, these private landowners wanted to preserve it as a permanent asset.

Wetlands

As the drive for nature protection proceeded, many natural environments once thought of in negative terms now were seen in a positive light. Mountains, rivers, and unproductive farmlands formerly bypassed as useless and unwanted were now prized as environmental resources. Wetlands were long thought to be fit only to be drained for productive farmland. After the 1950s, however, their value as habitats for plants and animals as well as for flood protection became recognized. Now one tracked the "loss" of wetlands rather than looked with pride on the number of acres drained, with the clear meaning that the decline was a tragedy for the nation. A marked change in terminology occurred as the more negative "swamp" and "bog" were replaced by the positive "wetland."

Natural Areas

Some of the first state "natural area" programs were formed in response to the new interest in preserving both unique geological features and unique flora and fauna. Land was inventoried for its singular qualities, and then funds were raised to acquire the more important properties. As time went on, this venture gained momentum from the unique partnership that developed between the state natural-area agencies and the Nature Conservancy, a national private conservation organization. State branches of the Nature Conservancy formed close working relationships with the state agencies to identify, acquire, protect, and manage natural areas. Citizens called "land stewards" often provided volunteer support to manage the lands.

Public Land Acquisition

A major component of the continuing interest in nature was the broad popular support for using public funds to acquire lands for permanent protection. The outburst of interest in outdoor recreation in the 1950s led to the creation in 1964 of the federal Land and Water Conservation Fund, whose monies were derived from the lease of federal offshore oil lands. After twenty years of vigorous acquisition activity from that fund, allocated for both state and federal purchases, Congress's interest in this strategy declined and it absorbed the funds into the federal budget to reduce the deficit rather than continue their allocation for land purchases. This decline prompted an increase in state funding sources such as bond issues, state sales taxes, real estate transfer taxes, lotteries, or revenue from resource extraction on state lands.

PUBLIC LANDS NATURE PROTECTION

The expanded interest in nature, and especially the role of land as habitat for wild plants and animals, had a distinct impact on the large federal landholdings of the National Park Service, the U.S. Fish and Wildlife Service, the U.S. Forest Service, the Bureau of Land Management, and the Department of Defense. When those lands had first been either reserved from sale or acquired, the objectives for which they were purchased and later managed had varied considerably. Each agency's lands now came under considerable pressure from those interested in nature, and their uses underwent review, based largely on the problem of adjusting older management objectives to a newer focus on natural values.

Parks

The national park system was the most amenable to new nature objectives. From its beginnings it had stressed preservation of natural beauty as a major purpose, and it always generated an interest in natural history in each of its park units. In the 1930s, the park system sought to bring a stronger wildlife focus to its program. After World War II it took up this objective more extensively, and under the influence of ecological scientists, developed an interest in restoring ecosystems that had prevailed prior to the advent of white European settlement. Many debates ensued over just what those "original" ecosystems were like, but the restoration plans usually involved an attempt to eliminate introduced species and to foster species that were indigenous to an ecosystem, such as the grizzly bear and the wolf in Yellowstone National Park.

The national parks were also prime laboratories for attempting to work out the appropriate balance between natural ecosystems and human pressures on them. The problem of that balance was inherent in the National Park Act of 1916, which required that the natural resources be protected and that human use and enjoyment of them be promoted. Since human use reached new levels each decade, the Park Service was preoccupied continually with the task of identifying the environmental impacts of park visitors and the "limits of acceptable change."

Wildlife Refuges

The system of lands known as the National Wildlife Refuges involved both land-based refuges, mainly in the west, and water-based wetlands for waterfowl along the nation's four north-south flyways. These were managed as habitat for plant and animal species rather than as locations for resource extraction. In the west, they were reserved explicitly from areas used for forage by livestock so that animals such as elk, antelope, or bighorn sheep would be able to thrive. Equally habitat oriented, and in a more explicit fashion, were the wetlands established to protect game birds. Attempts to cope with declining numbers of ducks, for example, by limiting the number of birds a hunter could shoot, called "bag limits," did not seem to work and led to the notion that the key to maintaining wildfowl populations was habitat where they could breed, feed, and rest. The first such refuge was located along the Upper Mississippi River, but soon refuges were established along every major flyway through which the birds migrated from north to south and back again. Acquisition of these wildlife refuges was funded by the "duck stamp," a stamp on duck-hunting licenses that hunters purchased. Migration routes were interstate and even international, hence the program was administered by a federal agency, the U.S. Fish and Wildlife Service, and arose out of treaties with Canada and Mexico to protect the migratory habitats. By the late twentieth century the wildlife refuges were one of the largest systems of land protection in the United States.

Because of its broad-based habitat approach, the U.S. Fish and Wildlife Service was able to accommodate without much difficulty the new ecological management objectives. The key turning point was its involvement in the Endangered Species Program that arose in the 1960s, which focused initially on the large land vertebrates but by the 1974 revision of that program had brought into the purview of "threatened and endangered species" a wide range of flora and fauna, including invertebrates and plants. Despite the fact that the Service had little experience in botany—its expertise was largely in animals—and had to draw on the Smithsonian Institution for botanical knowledge, it was able to adjust the old to the new because of its existing focus on the positive value of the biological system and especially because of its focus on habitat.

Forests

By the end of the twentieth century, forested areas seemed to provide a special context for nature preservation. Wooded areas in more settled urban and rural regions and forested wildlands provided varied settings in which natural values came to be expressed. Intense interest developed in the ecological systems of older forests—which had developed their own distinctive natural features over the years, their own unique complex of plants, lichens, salamanders, and birds—and in the threats to natural systems that came from forest-management practices such as clearcutting, from dividing large intact forest areas into smaller fragments, or from air pollution that greatly modified soils and forest growth.

In contrast with parks and wildlife refuges, forest management experienced considerable internal conflict, as those committed to the agency's traditional wood-production emphasis vigorously resisted new ideas about the value of a diversely aged natural forest. Forest managers did not think in terms of wide-ranging vertebrates or their habitat, but instead focused more precisely on those tree species that could produce commercially valuable wood—a subject known as dendrology. Foresters organized their inventory and classification systems for basic resource knowledge around stands of trees, classified by species. Their concept of habitat was derived from this "stand" approach rather than something based on the ecological needs and relationships of diverse flora and fauna. As the new biodiversity and ecological initiatives impacted forest management, therefore, they led to intense conflict between old and new ways of thinking about forests.

Grazing Lands

Publicly owned grazing lands managed in the West by the federal Bureau of Land Management contained not only game animals that might compete with cattle for forage but also riparian areas along streams that served as habitat for many species, which grazing tended to destroy. The Bureau was heavily geared to serving the interests of the livestock and mineral industries and resisted mightily the allocation of lands and waters to wildlife, either game or nongame. Because the agency found the newer ideas about appropriate uses so different from their preference for commodity uses, they vigorously resisted any incursion into their traditional management objectives.

Debates resulting from this clash between old and new went on for decades. One involved the composition of the district grazing boards that allocated the use of the public range. The customary pattern was a fifteen-person board, of which one seat would be allocated to wildlife interests. Gradually, new ideas came into the Bureau of Land Management as recreation and ecological objectives became more legitimate management activities. Some inroads came through environmental impact analyses that the courts required. Others came through an interest within the Bureau to protect the biota of riparian areas from overgrazing. During the tenure of Bruce Babbitt as secretary of the interior during the Clinton administration, state grazing committees were established with broader representation and broader policy objectives. But the clash between the old and the new continued, and changes occurred only slowly.

THE POLITICS OF NATURE

By the end of the twentieth century the human interest in nature had become a significant social and political force. It was, of course, not without vigorous opposition from those involved with development objectives, who could hardly accept efforts to preserve nature as a legitimate public aspiration. At most, they tolerated natural values as political realities that had to be placated in order for more important developmental activities to proceed. Despite this reluctance to accept change, the advent of nature as a widespread human interest and political force is one of the more remarkable features of the environmental era.

Especially significant were the common experiences that lay behind and sustained the growth of nature objectives. One was the way in which many people organized their family lives, including raising children, around home and recreational activities that stressed nature: backyard bird watching and flower planting, family hiking and camping, the use of nearby parks as nature areas, and, as children grew in age, more vigorous outdoor recreation. The various phases of this interest in nature can be tracked from the earliest exposure of young children to nature, to the vigorous outdoor recreation of young adults, to the less strenuous search for nature in raising children, to the less physically demanding forms of nature appreciation in one's older years.

Running through all this was a heightened exposure to visual representations of nature in books and magazines, in nature photography, and on wildlife television programs. This exposure has a history all its own. One of the first magazines to showcase color nature photography was *Arizona Highways,* produced by the state of Arizona. Only gradually, however, did such positive views of nature invade the American experience. They took many forms, such as calendars of nature scenes, large coffee-table books of nature photography, special photographic books of each national park and, as time went on, of the national forests and wildlife refuges, and led to the gradual extension of natural visual appreciation from national to regional and local settings. Personal photography was a form of combining nature exploration, natural science, and nature appreciation. It was no wonder that organizations that sought to enlist the public in preserving a piece of land or water or a species of wildlife did so by exposing them to color photography that enlisted their aesthetic appreciation. Television wildlife programs sharpened this interest dramatically. At first those programs seemed to emphasize the large megafauna, but later they explored the intricate lives of microflora and fauna, from termites, spiders, and centipedes to plankton, coral reefs, and plant-pollinating insects.

These experiences involved a steady increase in activities to acquire knowledge about nature. In America, in contrast with England, interest in natural history had earlier not been a characteristic part of popular education in the years before World War II, but especially after the 1960s, natural history became an increasingly important feature of both formal childhood education and informal adult learning. The aesthetic visual appeal of nature and the intellectual understanding of it seemed to reinforce each other. Environmental education in primary and secondary school involved both nature appreciation and ecological science. Nature education centers arose under the auspices of many public and private agencies, each one providing opportunities for "environmental education" for young and old. Visitor centers at the national parks, in some national forests, and increasingly at wildlife refuges fostered scientific knowledge about ecosystems and geology.

This intellectual curiosity about nature underpinned a wide range of learning activities, so that one could speak of a "nature learning economy" in which people spent considerable time and effort finding out more about the way the natural world worked. Some of these were civic enterprises in which long-established institutions such as zoos, museums, aviaries, and botanical gardens modified their passive observational displays into active educational programs that sought to enhance the role of their particular institutions in both child and adult education. New museums, each with a combination of appreciative and learning dimensions, arose to fill in gaps in more traditional institutions. As time went on, a new form of combining nature appreciation and learning emerged as ecotourism, which combined travel to places of unique natural interest with learning more deeply about the natural world. Among the retired middle- and upper-middle classes, ecotourism was especially popular through such organizations as Elderhostel.

For some people, an interest in learning included action to protect nature. Visiting wetlands to observe, identify, chart, and monitor species and to study their flora and fauna reflected an intellectual interest in these natural areas, but it also gave rise to considerable public support for wetland preservation. Another example was the issue of endangered species, which focused interest on species that were rare, endangered, or of "special concern." Organizations arose to bring together aesthetic, learning, and research interests with financial support for programs to benefit individual species such as bears, wolves, raptors, bald eagles, loons, or butterflies.

The interest in nature often brought together younger and older generations with a common interest that cut across different ages. It brought the future into the present as adults and older people became interested in nature preservation as a way of providing something valuable for their children and grandchildren. People justified many preservation ventures on the grounds that they wished to "save" something for future generations. This attitude underlay the popularity of private land conservancies as objects of memorial gifts for family members and friends who had died. And it underlay a curious but instructive response of people to the question some economists posed as to how much people would pay to preserve land. Given three choices, "recreation value" (preservation for current recreation), "existence value" (preservation for nature), or "bequest value" (preservation for future generations), the last evoked the most positive response.

This varied set of interests in nature was particularly striking among urban and suburban families. This is not to say that everyone within cities maintained such objectives, for cities harbored a wide range of values, perceptions, and ideas. But the urban context was quite different from the rural context of earlier American life; cities harbored appreciative views of nature while rural areas continued to express extractive views derived from the past. The evolution of interest in nature is a part of the evolution of urban society. As the focus of American society shifted from an earlier rural past to an increasingly urban future, many changes took place in what people valued in their lives. The resulting environmental culture, which had deep roots in the nation's cities, spread to the entire society as an integral part of the spread of urban culture.

The interest in nature preservation was not an attempt to "return" to some imagined past but a resolve to create a more desirable present and future. It was the environment of urban life, not an imagined and longed-for pre-urban life, that nature was intended to enrich. Nature appreciation evolved within the modern urbanized community and marked a major departure from the more extractive values of a previously dominant rural society that looked upon nature as a commodity to be exploited rather than as a natural environment to be appreciated.

5 THE WEB OF LIFE

The public interest in nature, known widely as "nature preservation," had a close but distinctive counterpart in the search for a less contaminated or less polluted environment. Less polluted implicitly meant "more natural," a world that could experience less contamination from human activity. In the heat of public debate, absolutist terms often took over, thus more "natural" often was described as "pristine" in the same sense as "Eden" constituted the absolutist form of "nature preservation." But more proximate or relative terms described more accurately the direction of environmental objectives as efforts to reduce pollution continued and as environmental action moved from one stage to another toward a "cleaner" environment.

While much interest in pollution concerned human health, it was not wholly divorced from conserving the natural world. Water pollution—and air pollution because of deposition from the atmosphere—involved aquatic life in lakes, streams, and oceans. As they circulated through air, water, and land, persistent, nonbiodegradable toxic chemicals made little distinction between humans and the wider environment in their dispersal and effects. Often the formulation and administration of programs for environmental improvement, and the corresponding competition for funds among them, led to shifting emphases, now on human health, and then on the broader environment. Scientific knowledge about the way the environment worked seemed to bring facets of the natural world together as the effects on humans and the wider environment were found to be closely interconnected within a mutually shared medium of air, land, and water. Science continually reinforced the sound-bite notion in John Muir's well-known phrase that "everything is hitched to everything else." The interest in pollution arose initially from the direct human experience of polluted air as smoky skies or polluted water that was unfit to drink or that killed fish. This human experience came to be defined in more systematic ways by ecologists, who spoke in terms of biogeochemical cycles and the human impact on natural cycles. Natural processes that cycled chemicals through the natural world came to be overloaded with additions from humans. Thus, too much sulfur discharged into the atmosphere from fossil fuel burning could fall as acid rain or snow or dry deposits, acidifying lakes and streams and corroding human-made structures. Often only "acid-loving" species could tolerate the changed circumstances. Thus, one spoke of the way in which human-created chemical discharges "overloaded" more natural cycles in such a way as to produce "kickbacks" in natural systems that people found to be unacceptable.

Pollution first became salient to residents of cities, where it was an integral part of urban congestion. Large numbers of people living in a small space gave rise to undesirable environmental circumstances affecting many. Smoke arising from heating many homes close to each other, combined with smoke from urban industrial activity, concentrated pollution that if more widely dispersed in the countryside aroused far less revulsion. Increasing amounts of waterborne waste from households and from commercial and industrial establishments led to demands that it be removed. Solid waste from a wide range of human activities prompted tactics to either reuse it, burn it to reduce its volume, or carry it away from the city. Few thought of these cases as problems in the relation between natural cycles and human overloads; they were regarded simply as unpleasant or unhealthy circumstances that reduced the quality of urban living.

By the mid-twentieth century, these facets of pollution had become more extensive. Urban households and industries located in urban areas discharged increasing amounts of sewage and industrial wastewater into nearby rivers and lakes with limited capacity to absorb the waste. Processes of biological decay or biodegradation now became overloaded with waste that consumed oxygen in the water to such an extent that it was not available for aquatic life, which, in turn, died. Solid and chemical waste was transported to nearby rural areas, which became dumping grounds for waste of urban origin, creating hostility and opposition from rural people who sought to ward off this potential harm. In addition, air pollution was diverted from the communities surrounding industrial plants by erecting tall smokestacks that directed it "away" to contaminate land and water downwind; during transport over long distances, moreover, it was often transformed into even more harmful chemicals.

In ways such as these, pollution was not confined to cities but was widely spread throughout regions, whole continents, and even the entire world. This linked sources and effects over long distances and identified more sharply the difference between natural chemical cycles and their overloading through the increased amount of human-made chemicals. One of the most dramatic examples was lead, a mineral associated with many soils and rock that underwent slow and natural weathering, but that increased markedly in the atmosphere as it came to be used as a gasoline additive and was spread by gasoline motors over wide areas. By the 1970s higher blood-lead levels in humans had spread throughout the world, with the heaviest concentrations occurring in the more industrialized countries and lesser ones in nations such as Nepal that had experienced little industrialization.

The increase and spread of pollution through air, land, and water, collectively known as a multimedia complex of environmental circumstances, reinforced the experience of a web of natural processes that were interrelated. The perception of a "web of life" emerged in how pollution was thought about and came to make increasing sense to those who followed the trail of chemicals in the environment. As long as pollution was confined simply to unpleasant aspects of urban living, the experience of the "web of life" was rather limited, but as the reach of pollution spread over larger geographical areas, it generated a sense of comprehensive contamination. Fishkills in lakes and rivers, contaminated underground drinking-water supplies, or dying trees injured by air pollutants all enhanced the sense of the complex environmental system itself and its workings when "overloaded" by human-caused pollution.

The environmental experience of the last half of the twentieth century tended to produce a rough link between the "nature" of "nature preservation" and the "nature" of a "more natural environment" less contaminated by human-created pollution. In the world of specialized science, policy, and administration, "nature protection" and pollution control remained somewhat separate. But those responsible for the management of land and water found themselves continually pressed to foster a more integrated science and policy that brought the two together. In the late 1960s and early 1970s, major efforts were made to affirm their close connection, especially in state administrative arrangements to bring conservation and environmental protection together in a single agency. Even when that did not occur, scientific knowledge continually revealed in ever more detail the way in which the two worlds of natural resource conservation and environmental protection remained closely interrelated.

PENETRATING THE BIOGEOCHEMICAL WORLD

From Experience to Knowledge

Pollution became a forceful presence in the minds of people through personal experience. In its earliest form in cities, such as smoke or noise from a nearby factory, smells from domestic animals kept by urban people or from rendering plants processing animal products, and waste dumped into streams that "fouled" one's community or home, it was known as a nuisance. From these direct human experiences, pollution extended into matters of personal health and recreational enjoyment as the environmental causes of disease became more obvious and rivers became unsuitable for swimming and fishing. This personal experience, then, led to systematic investigation, which in turn gave rise to much more environmental knowledge about biogeochemical cycles that linked the personal and the immediate with the workings of the biosphere upon which human society depended.

The Dimensions of Pollution

The first dimensions of pollution were simple matters of obvious air and water contamination and solid waste. Over time, knowledge about the sources of pollution increased to reveal more complexity. Each resulting action was relatively limited, giving rise to laws that constituted an initial step toward coping with pollution, but such action equally gave rise to increasing knowledge that broadened the scope and range of potential further action.

As public support for dealing with pollution problems grew in the years after World War II, demands for action focused on all levels of government: local, state, and national. But some of these were more responsive than others. Local governments had few required technical and administrative skills essential for action, and when, for example, cities tried to control air pollution, they were often challenged by industries with superior scientific and technical resources, by means of which they could ward off action through the courts. State governments had greater resources than cities, but their weakness was reflected in the degree to which they relied on representatives of industrial sources of pollution to shape and administer their programs. Hence, as the beginnings of air and water pollution control emerged, demands escalated for federal initiatives to require states to act. This resulted in a series of laws in the 1960s, culminating in the Clean Air Act of 1970 and the Clean Water Act of 1972.

These beginnings at control were expanded through additional laws such as the Safe Drinking Water Act of 1973 or the Toxic Substances Control Act of 1976 aimed at preventing harmful chemicals from being injected into the environment in the first place. Solid waste was left primarily to state and local action, since it was thought of more as a "local" or "regional" rather than an interstate phenomenon, as was the case with air and water. This gave rise to a wide range of local and state strategies to manage waste: incineration, recycling, and "sanitary" landfills. Only when states sought to restrict the import of solid waste from other states did it appear that some sort of federal action regarding solid waste was appropriate.

A major change in environmental pollution came about with the rapid growth of concern about toxic chemicals. These were different from "conventional" pollutants because they were not biodegradable, that is, they did not change from harmful to relatively benign chemicals through natural decay, but were long-lived and persisted in the environment. There were, of course, toxins or "poisons" in the natural world, but increasing attention was given to synthetic or manufactured chemicals that were produced precisely because they were toxic, such as pesticides or herbicides, or because they were desired in industrial applications because they did not degrade. There was no question that very high levels of exposure to these chemicals were harmful to both humans and the wider biological world of plants and animals. But with time, concern also arose for the possible harm of persistent, low-level exposures to synthetic chemicals such as DDT and PCBs that were found both in humans and in the environment throughout the world.

The possible harmful role of toxic chemicals was well recognized by scientists even during World War II, when DDT was used widely by the armed forces against body lice and other organisms injurious to humans. With the publication of Rachel Carson's *Silent Spring* in 1962 and the later administrative proceedings in Wisconsin concerning a possible ban on the use of DDT, the literate public learned of the problem. But the potential harm became even more sharply focused through the effects on workers in plants that manufactured the chemicals. One of the first cases to come to the public's attention was that of kepone. Manufactured in Virginia, kepone was a synthetic chemical used to destroy the nervous systems of insect pests, but it also turned out to have marked effects on the nervous systems of exposed workers. Even more widespread in its potential impact was the proliferation of industrial waste-disposal sites for toxic chemicals throughout the country. The initial widely known case involved a site at Niagara Falls, New York—Love Canal—in which citizens in a community surrounding a toxic-waste-disposal site documented the adverse health effects for those living near the site. The resulting controversy led to a massive cleanup project at Love Canal and the government purchase of homes of those living there.

The presence of toxic chemicals in the environment gave rise to a new perception of pollution as an even more wide-ranging problem, extending far beyond smoke, overloaded landfills, or rivers that were not swimmable or fishable. That perception had four dimensions. First, toxic chemicals were persistent; they did not readily degrade in the environment but lasted for a long time. Second, they were ubiquitous; they extended to communities, regions, nations, and the world, and they seemed to travel by air as well as water from sources to effects, and could be found everywhere, from the fat of penguins in Antarctica to the bark of trees throughout the world. They were so widespread, in fact, that so-called control groups, people who had not been exposed, often could not be found to make comparative epidemiological studies. Third, they were mysterious because they penetrated both humans and the environment long before they were identified and measured, and they appeared almost "accidentally" rather than as a result of systematic observation. And fourth, they seemed to be completely uncontrolled; they had become diffused throughout the world and had accumulated without much effort to prevent their spread and accretion. These four dimensions of toxic chemicals—persistence, ubiquity, mystery, and lack of control—shaped the perception of toxic chemicals in general and underlay the response to each new specific case.

Pollution issues involved three major types of processes and hence three types of knowledge: sources; transmission through air, water, and land; and effects. As the scope and range of pollution increased, scientists tackled each of these dimensions. Knowledge about sources was the most extensive. In fact, as the debate over control of pollution proceeded, much of the argument came to focus not on point of origin, but on the relative responsibility of the different sources, each of which sought to shift responsibility to others. Those who sought to keep an even hand on the causes of pollution frequently found themselves buffeted among the different sources.

Although sources of pollution were readily identifiable, the transmission of chemical pollutants from their sources to their destinations was not. Knowledge about transmission was skimpy, and pieces of the process only came to be known as funds slowly became available to conduct research. For many years sulfur dioxide was considered local in its effect, and tall stacks were used to divert emissions away from local communities. But when sulfur emissions were tracked through the atmosphere from sources to destinations downwind, these measurements resulted in the knowledge that emissions not only were transmitted over long distances but also were transformed in the process, in this case from sulfur dioxide to sulfuric acid or "acid rain."

Means of transport came to be an integral and extremely difficult feature of toxic chemicals. Chlorinated hydrocarbons in the Great Lakes, for example, were found to come primarily from the air rather than from the land. Toxic chemicals were found in remote areas throughout the world such as Antarctica, where air transport and the food chain were thought to be major sources. Movement of volatile organic compounds was even more difficult to trace because as they were transported they tended to be deposited and then to revolatilize again and again each day, requiring continuous twenty-four-hour measurement, thereby increasing the cost of tracking them. As these chemicals worked their way up the food chain, they spread even further and became more concentrated in body fat with each step.

The effects of chemicals were difficult to examine, since they seemed to be influential at very low concentrations. For scientists, this meant measuring them at parts per million, the standard concentration used in the late 1960s, but this became parts per billion, then trillion, and then quadrillion in the 1990s. The high cost of such measurements meant that chemical knowledge was skimpy, and funds for measurement were available only in a few targeted cases. By the late 1990s knowledge about the role of small particles, or particulates, in air pollution had shifted from the large particles in smoke, to particles of 10 microns in size, then to particles of 2 or 3 microns, largely on the grounds that it was the smaller particles that penetrated deeply into the lungs and caused most of the health problems. But for the most part the effects of chemicals at low levels of exposure remained rather uncharted, always leaving a sense of mystery about them.

The Increasing Scope of Pollution

Over time the scope of environmental pollution issues broadened widely as environmental science outlined and then detailed the way in which natural biogeochemical cycles carried their pollution loads from site-specific sources to far-ranging destinations and effects. This led many environmental issues to shift from local to regional to national and international dimensions in defining the way in which environmental pollution worked. In response, research and technology on the one hand and policy and action on the other shifted equally in scope. The "environmental problem" no longer consisted only of what harm an industrial plant might create for one's neighborhood, but now included the entire globe, with worldwide consequences.

This perspective first appeared in reaction to the radioactive fallout from nuclear tests and the atomic bomb explosions of World War II. Radioactivity could be measured in clearly defined atmospheric pathways from the sources of the explosions and tracked as the radioactive materials circulated several times around the globe. Some of the first popular as well as scientific understanding of long-distance transmission of harmful chemicals came in the 1960s when a scientist at Washington University in St. Louis, Barry Commoner, circulated such information from his organization, the Committee on Nuclear Responsibility. Not long thereafter, a scientist at Pennsylvania State University, John George, identified DDT in the fat of penguins in the Antarctic, thereby confirming long-distance transmission as the only feasible explanation as to how this chemical, purely human and not natural in origin, could appear there.

The identification of such pathways in the transport of chemicals, linking sources and effects at great distances, came slowly and often only in response to specific problems that a particular scientist took up as his or her project. Only gradually did such information jell into a focused subject in its own right, and even then one could speak only of a fragile and partial focus. The cost of pathway research and the enormous debate that it generated greatly slowed its development.

In the case of lead, the research did not actually track the metal as it moved within the atmosphere but, like the DDT research, identified its presence so widely in soil, in humans, and in such diverse places as to infer movement with great credibility. Clair Patterson, a geophysicist with funds from the Atomic Energy Commission, became interested in the conventional assumption that human exposure to lead came from weathering rock, which released lead into the soil, which was then taken up by plants, and then led to human exposure through eating. Patterson measured lead in many media: headwaters of streams and their estuaries; the surface, middle, and deeper levels of the ocean; deposits in Greenland ice; human blood from cities and from remote areas in Nepal; mummies of "ancients"; and in research laboratories contaminated with lead from the air and human contact. These measurements identified wide variations in the presence of lead that, in turn, pinpointed the source as lead in gasoline that cars spewed out with their exhaust.

Nitrogen oxides and sulfur dioxide emissions from automobiles and trucks on the one hand and from coal-burning utilities and fossil-fuel-burning industries on the other were found to be sources of pollution at places far distant from their locations. As these emissions traveled through the air in the presence of sunlight they became oxidized into sulfates and nitrates that fell as harmful acid rain, snow, fog, or dry particulates. Using balloons and aircraft to follow the air plumes from their sources, scientists produced precise measurements that established the entire process. Analysis of the chemical composition of both wet and dry atmospheric deposition identified air as a major source of water contamination. The issue was raised especially by those concerned with water quality in the Great Lakes and in Chesapeake Bay. The focus on long-distance transport was enhanced significantly by the ozone problem in the northeastern part of the country as downwind states in New York and New England complained that sources far upwind in the Midwest caused their ozone problem.

In the 1980s scientists began to think that air pollution was changing the Earth's climate and atmosphere as a whole. Two issues became especially significant: one involved the effect of chemicals from human sources on the upper ozone layer, which created an "ozone hole" and increased human exposure to harmful ultraviolet radiation. The other was "global warming," in which increasing amounts of carbon dioxide from fossil fuel and other human-made chemicals trapped more heat in the Earth's atmosphere, leading to a slow increase in global temperatures and possibly to an increased frequency of violent storms. The relevant atmospheric science was part of a growing activity known as climate research, which had been slowly evolving since World War II. Progress in such research depended upon the development of precise measurement techniques that were reliable and consistent in the varying circumstances of atmospheric conditions. At first, stratospheric ozone measurements came from on-the-ground monitors, but they became far more convincing when taken by aircraft that maneuvered in and out of the ozone layer and took direct measurements. As this knowledge grew, it greatly extended the perception, both popular and scientific, as to the diverse ways in which human activities affected the environment of the Earth in its entirety.

Although knowledge about waste transported through the air greatly extended perception of the area impacted by pollution, so also did knowledge about solid and chemical waste transported by land. In this case, the changing perspective was due not so much to scientific knowledge as to the waste-disposal industry and the constant temptation to go farther and farther afield to disposal sites. Cities and industries were the main sources of waste, the unused by-products of manufacturing and consumption. In earlier years it was disposed of in nearby dumps, by recycling, or through incineration, but incineration only increased air pollution, and dumps were usually unsupervised and could readily lead to contaminated groundwater. Seacoast areas often sought to dispose of their waste in the ocean, resulting in waste-strewn beaches when tides washed waste material back onto shore. Again and again, the waste problem seemed to be intractable, driving continuous attempts to discover places far beyond the waste's origin for disposal.

Within the United States, some areas became favorite targets for disposal sites, usually communities of lower income and with fewer abilities to engage in political protest. But contrary to expectations, these communities did resist, mounting protest after protest against waste-disposal siting in their "backyards" and also giving rise to a generalized "environmental justice" movement. Middle-income and more affluent communities remained rather unaffected by the threat of waste-disposal sites because of their greater potential political clout.

Some states came to be disproportionate producers of waste and others disproportionate recipients of disposal sites, and this situation introduced long-distance transport of waste, which expanded the waste-disposal problem to an interstate dimension. The same type of distinction arose on the international scene, as the industrial nations found opportunities for disposal in third-world countries, for example in Africa. This, in turn, created a sharply defined issue between the two areas of the world in which third-world countries demanded that such disposal be conditioned on prior approval of the receiving country. The issue became a focal point of debate in the United Nations Environmental Programme and led to the international Basel Convention to regulate the practice. The debate sharpened the differences between the developed and the undeveloped nations and helped to identify the way in which seemingly local environmental circumstances now had become worldwide.

For many Americans, the issues surrounding nuclear generation of electric power brought together rather sharply many of the dangers of the "chemical age." In the early 1970s, nuclear accidents came to define the hazards of nuclear energy, and these were deeply ingrained in public experience by the accidents that took place at Three Mile Island in the United States in 1979 and at Chernobyl in the Soviet Union in 1986. Public confidence in those who promoted nuclear energy steadily declined. The Atomic Energy Commission contributed much to this loss of confidence when it declined to consider in its siting decisions the adverse environmental effects of its technology. Its actions became one of the earliest cases of citizen participation in public decisions through the application of the 1969 National Environmental Policy Act and the activity of "whistle-blower" nuclear engineers who worked with citizen groups in challenging the wisdom of the commission. In the ensuing years, the permanent disposal of radioactive waste predominated as the primary legacy of the nuclear issue. The nuclear "establishment" continually argued that they could "solve" the waste problem, but its record undermined public confidence in its assertions. Amid all this lingering dispute, the construction of new nuclear power plants sharply declined.

How Clean Is Clean?

Specific target levels for pollution reduction were the fundamental feature of control strategies. How clean should the environment be? The world of environmental affairs was dominated by management and debates over implementation strategies. But within those debates arose a more fundamental debate over objectives. Just how much cleanup? Just what quality of environment should be the goal? Environmental organizations and the environmentally conscious public whose values they expressed fostered progressive improvement in environmental quality, much as those interested in economic growth encouraged increased production, consumption, and employment. But the environmental opposition questioned and undermined the wisdom of environmental goals, thereby serving as a major restraint on progress in environmental quality.

Initial standards in air pollution came from revulsion against smoke in cities. There the goal that shaped action was to reduce the smoke that impaired visibility and dirtied houses and clothing. The benefits in cleaner air that came from first smokeless coal and then natural gas heating spoke for themselves and involved little in the way of more precise determinations of the amount of cleanup desired. As environmental knowledge identified the wider effects of air pollution on vegetation, buildings, and human health, determination of the level of cleanup became more complex. In 1963 Congress directed the U.S. Public Health Service to develop "criteria documents" in which the service would summarize the scientific data and establish a specific goal for the desired air quality for the community. Thus began an intense, persistent, and never-ending battle over "how clean is clean," with the environmental community desiring higher levels of air quality while the sources of pollution resisted them, arguing that such levels were not necessary and too expensive.

The argument soon took the form of asking what the "benefits" of cleaner air were and of a debate over the validity of scientific data concerning the harmful effects of air pollution. To the general public, the technical context of the arguments became difficult to follow. While the public could easily understand what level of air quality was at stake with regard to visible smoke, much of the effect of air pollution on vegetation, materials, and human health was more difficult to perceive and understand. As the debate proceeded, the varied effects that were thought to be undesirable became more sharply etched: rivers and lakes where acid rain had reduced fishing should be restored; air pollution that caused health problems and premature death should be reduced; areas such as parks and forests that were prized for the clarity of their air should not be permitted to undergo reduced visibility. Each such case continued and expanded the issues involved in "how clean is clean" air quality.

And so it was also with water quality. The debate over water pollution in the years prior to the federal Clean Water Act of 1972 led to major decisions as to how clean streams should be. One was the rejection, after long years of dispute, of the idea that rivers should serve as places to dispose of waste. This role of rivers came under considerable fire in the 1950s and 1960s and gave way to other objectives. The goal identified in the 1972 act was that all water bodies should be made "fishable and swimmable," that is, of a quality to enable aquatic life to thrive and to permit human contact.

But, as was the case with air quality, such ideas did not resolve the continuing problem of just how "clean" streams should be, so the debate continued. What level of water quality was required for what kind of aquatic life? What level of water pollution was harmful for human swimming and recreation? Despite dispute over details, the Clean Water Act of 1972 marked a major change reflected in the act's definition of its goal: that all aquatic ecosystems should be maintained in their "ecological integrity."

Solid waste disposal presented its own problem of standards. The initial impulse that continued to drive thinking about waste disposal was to "get rid" of it in some fashion. Early in the evolution of urban waste management, efforts were made to recycle it through the work of scavengers or waste pickers who removed articles from waste piles to sell for reuse. As the amount of waste grew, the idea of burning it came to be popular. But the resulting air pollution spurred changes in this practice in the 1960s, as both municipal dumps that were burned regularly and backyard trash burning were rejected. This launched the search for waste-disposal sites in rural areas outside cities, which in turn created tense controversy between the cities and the countryside: the cities argued that such sites were a necessity for them, and rural areas objected to use of their communities as waste dumps. It also led to intense interest in the cities in recycling commercial and household waste such as bottles, glass, plastic, metal, and paper. Recycling gradually evolved into a new "ethic" that pervaded many sectors of society.

From this experience there evolved a hierarchy of alternatives that came to underlie thinking and policy about waste disposal. It ranged through five priorities: (1) reduce the initial use of resources to produce less waste; (2) use production processes that would produce less waste; (3) recycle waste; (4) incinerate waste; (5) dispose of the residue in landfills. It was more than difficult to translate this fivefold sequence of desirable waste policy into practice. But it provided a framework within which the general notion of "how much waste" and "what to do with it" could be worked out in practice. Among all these alternatives, the notion of "pollution prevention" was the most popular idea. But this also was subject to considerable controversy. Those who produced waste argued that recycling was pollution prevention, while environmentalists argued that this made a mockery of the priority system. "Real" pollution prevention involved steps one and two, in which the burden would be on industries to produce less waste during production.

The notion of "how clean is clean" was sharpened in the drive to clean up or remediate old waste sites and especially hazardous waste sites. Programs emerged at the federal and state levels not only to restrain the production and disposal of new hazardous waste but also to clean up previously contaminated sites. The problem was enormous. From the earliest days of industrial manufacturing most waste had simply been discarded in the most convenient way, on-site or nearby in wells or in streams. Chemical waste had since leached into underground water supplies. As attempts to reduce continuing hazards from past practices proceeded, the central question came to be "how clean" should the cleanup operation be? Which techniques should be used: volatilize contaminants in the air to reduce their level in land or remove contaminated soil and cart it away to be disposed of elsewhere? Affected communities demanded a higher level of cleanup, and the sources resisted such demands and argued that the harm was illusory. Management agencies sought to mediate between such pressures and to justify levels of cleanup in terms of some health statistic that defined the potential harm from the contemplated future use of the site.

Intruding with considerable power into this problem of defining levels of acceptable contamination was the role of toxic pollutants. Strategies to reduce toxic chemicals to harmless levels had not been found. Scientists had worked on many fronts on this problem and met with limited success. In some cases, such as oil residues, bacterial agents had been used to decompose them—a process called "bioremediation." But most toxic chemicals seemed to defy such approaches, and the usual method was simply to remove the chemicals to other locations, or even to cap the areas and build fences around them in the manner of a permanent waste mausoleum. Such limitations to environmental technologies put considerable pressure on the topmost priorities in waste disposal, resource use reduction and manufacturing processes that produced less waste, especially toxic waste.

The task of defining limits of pollution or standards of cleanup became more complex as environmental science revealed the complexity of environmental contamination. On the one hand, knowledge revealed wide-ranging effects on both humans and the environment. The range of health effects expanded from more acute and chronic disease to cancer and then to a wide range of developmental problems such as impaired reproduction, abnormalities in fetal development, stunted neurological development, impaired immune systems, and reduced lung capacity. The focus on harm to humans expanded from healthy male workers to include the entire community of children, women, and the elderly. The issue also had significant ecological implications, as study after study added to the complex of knowledge about the effects of pollutants on rivers and lakes, soil, crops, forests, the larger wildlife, and the entire range of flora and fauna.

A significant dimension of "how clean is clean" was action to prevent cleaner environments from becoming more polluted. This became known as the non-degradation problem, and it became especially important in air- and water-quality issues. The question arose first in water quality. Federal officials discussed among themselves the very commonsense question: should some waters be allowed to become dirtier while efforts were made to clean up others? To permit the first was to make no overall progress in cleanup efforts. Hence in 1968 the Department of the Interior developed regulations to prohibit degradation of water quality. The proposal aroused much controversy, led to "antidegradation" requirements in the Clean Water Act of 1972 and to even more precise ones in the 1977 revision of the act. In the ensuing years, environmentalists sought to retain a focus on this issue amid intense resistance from sources of pollution as well as state agencies, and due to their actions, the Environmental Protection Agency issued relevant regulations that required states to classify waters and identify those of higher water quality that should be protected against degradation.

A similar set of problems arose for air pollution. In response to initiatives from environmentalists, the courts accepted the argument that the Clean Air Act of 1970 required the Environmental Protection Agency to develop a nondegradation strategy for air quality too. In response, the agency classified air into three levels of quality: Class I, Class II, and Class III, with Class I being the highest quality requiring the highest protection, and established processes to which governments had to adhere if such areas were subject to new sources of pollution. These provisions became more elaborate in the revision of the Clean Air Act in 1977 and became a central point of debate in air-quality affairs. Environmentalists sought to defend the law and the regulations, while developers sought ways of evading and modifying them.

The antidegradation debate in urban areas took on a different twist, requiring that if new development produced more pollution, other sources of pollution must at the same time be reduced. The issue then came to be: reduced by how much? How much of an offset would be required? Would the reduction be only 1:1, that is, one unit of reduction for each unit of increase, or would it be a greater ratio, such as 1:1.5, in order to improve air quality? Argument arose over not only the acceptable ratios but also the data on which they were based. How did one know that a given amount of existing pollution was really that specified amount, that the numbers were "real world" numbers in comparison with the "real world" numbers that reflected the new emissions? In the face of pressures from development, represented in both private and public agencies, the prevention of degradation was as difficult to maintain as was progress in cleanup.

6 LAND DEVELOPMENT

One of the most persistent and intractable roadblocks to environmental objectives was land development. Developments of all kinds seemed to threaten to destroy natural habitat, to produce more pollution, and to make life increasingly more congested, impersonal, and fleeting. Open land was turned into housing, malls, or commercial and industrial establishments; colleges and universities expanded their campus facilities; more and more highways were built, adding more and more cars. The environment of human life was less and less open and expansive and more and more closed and confining. It seemed that massive forces were at work in a dynamic that threatened to engulf every piece of the environment in its orbit. And in the process, environmental values that people sought to expand were threatened with confinement, distortion, and even extinction.

People who sought to improve their environment found this set of circumstances to be almost impossible to modify. They could find ways to increase and protect nature amid a developing society so long as it was distant and could be isolated from the pressures of development. They could find ways to control some pollution, especially if it could be pinpointed to specific sources. But efforts to control development met with far less success. Often the focal point was the local community, the village or town, the city or countryside around one's home, where many an individual or community faced demands for more intensive development that threatened to radically change one's community and one's way of life. The most frustrating aspect of the demands was that they were aided and abetted by community leaders who held out the vision of the benefits that more development would bring. These leaders represented both those who would profit personally from development and those who sought to define their civic and government roles as development promoters.

Another factor was the persistent experience within the community itself of those who found development to their liking, who wanted its material and social benefits despite its adverse environmental effects. These individuals were caught up in the more highly developed society, with its dependence on the automobile, its greater freedom of choice, its higher material standard of living. Faced with these benefits, they had difficulty giving higher priority to environmental objectives. Hence communities were divided over such issues. For those who wanted to expand environmental benefits, the threat from development was more than exasperating. The situation presented a massive paradox. Although the environmental era witnessed some progress in realizing environmental objectives, the same decades witnessed the most rapid pace of development and population growth in the nation's history. The development juggernaut continually loomed large, thwarting efforts for environmental progress.

THE DYNAMICS OF LAND DEVELOPMENT

Complex forces combined to foster and sustain the drive for development. The most obvious was, of course, the developers themselves, including residential, commercial, and industrial developers, who sought to turn land of lower monetary value into a higher one by erecting buildings on it. Then there were those who benefited from enterprises associated with development, those fostering transportation facilities, especially the gasoline-powered car and truck, and the varied utility companies—telephone, electricity, water supply, and sewage—called for by more population. And behind it all were the vast number of businesses that sought a larger market for their consumer products. Their community organizations, especially the local chambers of commerce, fostered development in which all their members would find greater financial benefits.

Closely associated with these private entrepreneurs were local governments, which found in development the tax resources to conduct their public business. Early in the history of cities municipal governments found themselves involved in providing services—streets, water supply, sewage, fire and police, jails, education—and as time went on, more new services, such as public health, hospitals, environmental programs, parks and playgrounds, and libraries, were added to this burden. Each of these was paid for by taxes, and the main source of funds for most local governments was the property tax. Each government was caught between the desire to provide more services and the reality of limited income. They could raise taxes, but given the unpopularity of that strategy, they soon found that tax income would rise anyway with higher assessments from rising land values and took this route to expand tax revenue. This brought together the interests of private developers with the interests of municipal governments.

Governments were attracted to the increased revenue that would come from development and paid little attention to the resulting demands for more public services, adding costs to taxpayers. One strategy that had long been employed by municipalities was to require those who benefited directly to form special tax-and-spend districts to provide services. This was the case for services called for by the business community and led to special districts to provide them. Another strategy was to require abutting property owners to approve and pay for paving streets in front of their houses. But such strategies were gradually dropped in the face of arguments that most services were used by all and should therefore be paid for with taxes imposed upon all city residents. Property owners often objected to paving streets if they would have to pay the costs; to overcome this opposition, traffic engineers successfully advocated centrally directed street improvements paid for by the city's general fund.

As city services increased over the years, so did the level of city expenditures, thus a greater emphasis came to be placed on the added costs to the public with more development. Development would require more highways and streets, more schools, more fire and police, more libraries, and more parks. Such calculations also came to include the costs of more services for the elderly and the handicapped, more physical and mental health services, more environmental programs, more jails and programs to cope with crime and juvenile delinquency. Slowly but steadily one began to hear of the costs of development and not exclusively of its benefits. Consequently, in the environmental era some attention was paid by municipalities to the taxpayer costs as well as the taxpayer benefits of development. In some cases, the impetus for such thinking came from state governments, which faced their own share of the costs of development. In others, the impetus came from some environmental organizations that stressed the lower costs to the taxpayer of lower levels of development.

City governments were not the only focal points of development; private institutions were as well. Almost every private institution that provided public benefits, such as universities and colleges, museums, libraries, or hospitals, was driven to expand facilities, which meant development. Among these, the most pervasive were colleges and universities, which grew rapidly after World War II in terms of students, faculty, administrators, libraries, and research—all of which required more buildings using more space. In some cases, these pressures led to the relocation of entire campuses to the outskirts of a city; in others, it led to land-use conflicts between institutions and their neighbors; in still others, it led to buildings being placed on traditional campus greenspaces. Among students who became interested in environmental affairs in college, issues of campus greenspace often stirred them to action.

Land development was an integral feature of urbanization. It was in the city that construction raised land values and, in turn, spurred demands for social services financed through taxes. But the impetus behind development did not remain confined to the city itself. Cities reached out to the countryside and did so in various ways long before the appearance of twentieth-century suburbanization, which came to be called "urban sprawl." Cities established transportation links among themselves as promoters of economic growth found that much of the expanding economy was based on interurban commerce.

After World War II, moreover, the cities began to expand their influence not just by suburbanization, but also by urban people developing a marked interest in the wider countryside. They engaged in recreation there; they visited there as tourists; they found attractive pieces of land there that they bought for weekend recreation or permanent second homes. In doing so, they provided a new basis for the rural economy that came to be called the nonfarm economy. Rural towns became more stable economically, and each, in turn, developed its own visions of development. The dynamic was similar to the earlier history of cities: owners of real estate sought to obtain income from rising land values; local governments looked upon rising property values as a way to increase tax returns and thereby increase municipal budgets to provide for expanded municipal services. Federal and state governments contributed their own funds to this process, especially federal rural development agencies.

Tax benefits from development steered many municipalities into approving large commercial establishments such as shopping malls. Residential development was much less desirable, since it required significant public expenditures to pay for the accompanying utility and family-related services. In the language of municipal planners, residential developments were "losers." They required schools for the children, street maintenance from snow in the winter, and street repairs in the summer. In contrast, shopping complexes, light industry, and research labs were big "winners," "the cream of the ratables." Their environmental impact was fairly light and carried relatively high assessments, hence equally high property-tax returns. As long as the property tax was a major source of municipal income, municipalities enthusiastically joined with private entrepreneurs to sustain this kind of development.

It would be relatively easy to follow the environmental consequences of the development activities of builders, governments, and institutions. But one must also bring into the picture the desires of the general population as they sought out and accepted the material benefits of development. Promoters and producers were a big part of the development picture, but so were users and consumers. The automobile is deeply rooted in American society because it provides considerable choice and mobility that people want; the desire to improve one's standard of living by having a larger house with more space around it is a driving force behind housing development; students, those seeking medical treatment, and those using libraries desire and approve the "better" facilities that larger institutions provide. So there is a certain congruence between the values of developers and the values of users in sustaining the engines of development.

ISSUES IN LAND DEVELOPMENT

The continuous series of land-development issues that have arisen since World War II in community after community—in large cities, in small towns, and in rural areas—have given rise to a variety of attempts to cope with them in some fashion. One cannot say with confidence that they add up to a clear and effective strategy; they are more in the nature of separate programs, almost experiments. But they have promoted attempts to bring some features of development under control.

Most of the approaches involve, in one way or another, the role of planning in the use of land, devising strategies to modify individual cases of land development so as to arrive at what planners or the community think of as more desirable results. While developers and active segments of the communities in which they work are frequently at odds, some planners sit back, reflect on the whole picture, and come up with ideas as to how things can be done differently. Planning is a broad-ranging and varied activity, and much of it was shaped as federal funds for highways, housing, urban redevelopment, sewage facilities, and parkland purchases became available. Agencies distributing the funds required potential recipients to develop a plan as to how the funds would be spent and how the spending agencies could be held accountable. Environmental objectives became an ingredient in the planning process. Accordingly, planners began to work out details about open space, the protection of natural areas or wildlife corridors, floodplains and steep slopes, coastal areas and sand dunes, and wetlands—all of which came to be called "areas of critical environmental concern."

Environmental circumstances were a significant feature of zoning, the earliest form of urban planning. Zoning was an urban policy that separated different types of land into zones according to use so that one use would not interfere with another. Urban residents objected to nearby factories whose noise, odors, and air pollution affected their homes. As city dwellers sought to restrict industrial operations in their communities, the industries and commercial establishments, in turn, began to demand clearly defined areas that would be available to them. As a result, zoning identified three zones—industrial, commercial, and residential—to which those particular activities would be confined. Through this device, it was hoped that residents would have a major element of freedom from the environmentally degrading consequences of both industrial factories and large-scale commercial ventures. Zoning was also applied to the new suburbs on the outskirts of cities that grew up especially after World War II. One of its major features was density limits, a requirement that houses be built on lots of a certain minimum size. This followed what was thought to be a major objective in suburbanization, the establishment of homes with large areas of greenspace around them.

As environmental issues became a more explicit element in public policy, planners drew on the zoning experience and incorporated a wider range of environmental values in their ideas about land use. Many undesirable features of development could be avoided if growth were organized in a different way, and new ideas often came as a result of experience with earlier zoning decisions, which sometimes produced undesirable results such as suburban "sprawl." The innovation had two origins: one focused on the way in which development destroyed open space and on the concomitant search for a new pattern of development that would reduce this environmental harm. The other focused on the human relationships arising from the influence of the automobile and the possibility of promoting development that would foster more personal connections between people and their daily lives of home, work, play, and shopping.

Oregon was the first state in which new ideas about restricting sprawl and protecting open space through planning were put into play. In 1975 the state legislature approved a land-use planning system that emphasized preserving farmland by drawing a line around each town and city and requiring that development take place inside rather than outside that line. Within this set of guidelines, each municipality was required to come up with its own plan. The entire scheme was opposed by developers, who continued to seek to abolish it over the years, but it survived those assaults to bring about a different location of development in the state.

Another approach arose initially from development pressures in the Connecticut River Valley in western Massachusetts. It was fostered by the Center for Rural Development in Amherst, Mass., and it targeted one of the main unintended consequences of earlier zoning that had established minimum lot sizes in order to preserve open space. This had encouraged massive sprawl, and although it preserved space around individual homes, it also destroyed open space as a community asset. How to modify this? Their solution: require each large-scale builder to cluster the houses together and then establish a conservation easement on the remainder of the land so that it would be permanently devoted to open space. This would guarantee that open space would be protected from continual subdivision.

New ideas about how to plan buildings within the city arose as well. Here the environmental emphasis was less on open space and more on rearranging building so as foster community life on a more human scale: access to shops and recreation by walking rather than automobile, greenspace to break the overwhelming burden of buildings, and house design that would be more varied, less standardized, and more supportive of a relationship between indoor and outdoor activities. All this sought to reorder human relationships and to work open space into dominant conditions of congestion rather than, as was in the rural case, to work settlement into a dominant condition of natural space.

Land-use planners were much concerned about settlement patterns that were often at odds with the "forces of nature." Building in the floodplains of rivers invited destruction from floods that recurred at periodic intervals. Building on steep slopes invited landslides, including the destruction of houses, during times of heavy rains that caused soil instability. Building on sand dunes and in coastal zones put construction in jeopardy from shifting sands and periodic hurricanes that destroyed many homes. Such circumstances involved a similar problem: the location of development in places where the stability of the land was threatened by natural forces. After World War II they also involved a crucial element of public policy—disaster relief. After each such natural disaster, those who suffered property damage called upon the federal government for financial aid to rebuild. Legislators did not wish to be insensitive to the misfortunes of their constituents, but did it make sense for taxpayers to reimburse disaster victims only to encourage them to rebuild, risking still another round of federal disaster relief?

A set of ideas about such development was promoted by the landscape architect Ian McHarg in his book *Design with Nature* and focused on one goal: make sure that development avoids areas subject to natural disasters. Though this idea seemed sensible to the general public, it hardly made a dent in the temptations of developers to build in vulnerable places, and it especially bypassed the thinking of those who simply said that the floodplain was their home and they would return even though their house had been severely damaged. As population in given areas grew, the best building sites became occupied, so that population pressures were one factor in the development of disaster-prone areas.

Other approaches came to the fore. One was to require federal disaster insurance on the part of those who built in such places so that pressure on taxpayer-financed relief would be reduced. This did not have the desired effect: rather than reducing the drive to build in disaster-prone areas, making it less of a financial risk only made it seem more feasible. A second was to develop land-use plans that prohibited building on vulnerable areas such as floodplains, steep slopes, and coastal zones. In each case, the approved statutes affirmed the potential damages arising from such building and its fiscal burden on taxpayers; when landowners objected, the courts, in most cases, upheld the restrictions as a legitimate public purpose. Still a third approach involved programs to finance moving a town from a flood-plain to higher ground, a strategy that was used in a few cases as early in the 1930s and occasionally thereafter. The devastating Mississippi-Missouri River floods of 1993, however, moved this approach ahead considerably, as the cost of replacing the breached levees—structures on which farmers had long relied to protect their land—seemed unrealistic to refinance.

Attempts to incorporate environmental objectives into developmental practices invariably involved land-use planning. Although the focal point of this was local planning at the county and municipal levels, it often took the form of statewide land-use policies, such as in Oregon. In the early 1970s, efforts to establish general federal initiatives to promote such ventures failed, but a more specialized form of land-use planning dealing with the coastal zone did succeed. This was due to the interest of local coastal governments, all of which were engulfed by the pressures of increased population. Federal requirements for coastal planning were accompanied by attractive federal funds, and states were required to identify "areas of critical environmental concern" that required specialized management. Nevertheless, developmental forces usually dominated these programs; hence the environmental benefits were rather limited.

Development had important fiscal as well as environmental consequences, and municipal governments were often more impressed by the fiscal ones. This instigated a widespread move among local governments to require developers to absorb more of the costs of development—highways, schools, parks, and other public land uses. Florida was especially noteworthy for its requirement, not fully implemented, that local governments not approve development permits unless funds for the associated costs to local governments were approved and in hand. Washington State approved a similar type of program in 1992. This in turn led to intense reactions from developers, who argued that the program was an illegal "taking" of property without compensation, and the issue became a major force in the state's property rights movement.

The outcome of most of these issues depended on the attitudes of developers as they found that environmental objectives limited their freedom of action. Attempts to require developers to absorb the costs of development caused them to strike out with intense opposition. In some regions of the country where the issues had evolved over long periods of time, developers had come to accept the policy, but only partially. But in other regions, such governmental requirements were newer, and the opposition was more intense and dramatic. As environmental initiatives thwarted some specific projects, the development community tended to be somewhat more cooperative. Some came to terms with environmental advocates, urging a type of zoning that identified places where development could occur and places where environmental assets would be protected. At times this involved endangered-species issues when developers were stifled in their aims because of the presence of protected species. An approach known as "habitat conservation plans" evolved in which some accommodations took place, but often only at a high price demanded by developers.

ALTERNATIVES: LOCATION OR PACE OF GROWTH

These various alternatives for land development involved ideas about more environmentally acceptable ways in which development might occur. Development should take place here rather than there; it should be carried out so that it imposed fewer costs on taxpayers; it should be modified to be more environmentally acceptable. Such alternatives dominated most of the debate over development. Among such arguments were some attempts to focus on the larger issues raised by the clash between development and environmental objectives, but by the end of the century these were only suggestions that rarely obtained much support.

The most popular strategy was the attempt to conserve areas in which development would not occur. This might take the form of protecting open space not yet developed or of identifying areas with special plants and animals that should be protected. The entire idea of conservation as nature protection seemed to be fueled by the desire to create a significant alternative to the development behemoth. Attempts to control that behemoth were more than difficult; one had to rally large numbers of people in the community to shape local policies in the face of what seemed to be a powerful political combination of private developers, governmental leaders, and expansive institutions. In contrast, "land saving" seemed to be much more simple: identify land to be saved from development, then acquire it and protect it. One could readily join with like-minded people who, for the most part, agreed on both objectives and methods for reaching them. Moreover, one could draw upon latent public support for the idea that along with development there ought to be some land that was not developed. No wonder land conservancies had both a commitment to protect nature from development and a strong context of promoting public values.

The idea that valuable natural areas should be placed off-limits to development came from several sources. A significant impetus came from the Nature Conservancy's project to inventory areas for their unique natural qualities, such as the presence of rare and endangered species. As similar projects evolved in almost every state in the nation they helped to establish a firmer commitment to the conservation and protection of these natural areas, and that commitment was solidified even further by the way in which planners took up the idea of establishing programs to protect "areas of critical environmental concern" and worked it into almost every planning strategy. These programs were integral parts of federal planning, such as under the Coastal Management Act of 1972, as well as state land-use plans. Many states developed programs aimed at identifying and protecting such areas through voluntary action of developers to avoid them, cooperative and voluntary action with landowners to protect them, acquiring them as public assets, or restricting development on private property.

Most developers were unfriendly toward these activities; in the 1990s property-rights advocates came to look upon land conservation as a threat to private property. But in a few cases, they took a different tack. In the northern lower peninsula of Michigan, the Little Traverse Conservancy, which had established itself firmly as a beneficial community institution, began negotiations with developers to mutually respect each other's objectives by coming to some accommodations: the conservancy would accept some lands as more appropriate for development if the developers would accept some lands as more appropriate for conservation. Developers came to believe that this arrangement would enhance the values of their developments. This went one step beyond merely moving development from one place to another, to arguing that on some lands no development would take place because they were more valuable to the community for conservation purposes. An interesting twist to accommodation from the developer side occurred when a major real estate agency in this region of Michigan allowed its agents to give free memberships in the local conservancy to their clients.

There were some attempts, however, to define the value of such conserved land to the larger community by establishing general principles regarding the relationship between conservation and development. Few were successful, but even so they are instructive as proposed directions of thought. One proposal arose in Connecticut that the amount of land preserved from development each year should be equal to that devoted to development. This was quickly taken to be outlandish, and the episode only indicated how land preservation, though considered in a quiet way to be a community value, could not yet be elevated to an expressed principle or policy.

Nor could other alternatives that had a similar larger context. One, for example, was the proposal to establish development rights that could be transferred from one place to another so that development could be steered from environmentally more valuable sites to environmentally less valuable ones; this was known as transferable development rights (TDR). In this way the community could specify areas on which development should not occur and those on which it should occur. This followed the general practice of dealing with the impact of development by relocating it. But in several cases, an additional check was placed on TDR by establishing limits to the amount of such rights in a given area. The most extensive such case was the Pine Barrens of New Jersey. Here an overall limit was established as to the amount of future development that should be allowed for the Pinelands, rights to that development were distributed to all landowners, and locations identified as to where that development should and should not take place.

Environmental organizations also tried to deal with growth and development by proposing limits on the rate of new development. Such proposals occurred especially in areas that were growing rapidly and where the attendant problems were quite visible. There was no thought here of putting a cap on growth but only of reducing its rate, particularly by reducing the number of new building permits that would be allowed each year. One city that obtained considerable prominence for adopting such a policy was Boulder, Colorado, where the policy enjoyed strong support and where measures to preserve open space moved in tandem with support for slower growth rates. Although Boulder was able to exercise some restraints on growth, adjoining communities did not follow suit. One major result, therefore, was to direct growth to the area outside Boulder.

Interest in such policies was fostered in the early 1970s by the Zero Population Growth (ZPG) organization, which emphasized the role of population in development and early on considered the costs of growth in contrast with the benefits. It supported the new thinking that seemed to be taking place in cities like Boulder and generalized it to a larger question of the costs and benefits of urban growth. To what extent, ZPG asked, did the benefits of growth rise in comparison to the costs, and was there a point at which the costs to cities became greater that the benefits? Such thinking led to an economic study of cities that identified a tipping point at about 450,000 inhabitants, above which the costs of city services to residents grew steadily. Hence, the argument went, there was a point beyond which the economic benefits from urban growth were limited.

DEVELOPMENT AMID ENVIRONMENTAL PERSPECTIVES

Development pressures loomed large over the scene of environmental affairs. They seemed to threaten a wide range of environmental values and objectives, ranging from the protection of nature to the achievement of cleaner air and water and the development of more stable communities. They sharpened in one set of issues several of the more wide-ranging features of environmental affairs and hence held special significance for environmental politics.

Especially sharpened was the conflict between individual property rights and community welfare. Much of the growth of environmental awareness involved the widespread belief that property owners harmed the larger community by their decisions to develop and manage their property as they saw fit. While property owners asserted their inherent right to do that, the community often defined the issue quite differently in terms of the harm to the rights of the rest of the community. Development made that distinction more pronounced. Time and again widespread fear of and opposition to development was expressed on the grounds of its harmful consequences to the environment but also to the social fabric of human relationships and the community. This was a classic case of the conflict between the assertion of private rights and the assertion of the welfare of the public, which had long been the subject of both public debate and legal dispute ironed out by the common law.

In the midst of this conflict, some owners of private property held a different view. They were not so much interested in the profitable management of their land or in reaping its rising value, but in protecting it for its public or community value as natural land. This desire was shaped by several experiences. One was the way in which private ownership of natural lands had led the owners to appreciate them in their natural state and to hope future generations would do so as well. It was rather striking how much support some landowners, and the general public as well, gave to the idea that the value of their land was even greater for future generations than it was for its current use. Another factor, however, was the recognition that the land they owned would probably be sold for higher value, even for development, if it was passed on to children who now lived elsewhere and had no interest in owning it. These landowners often concluded that some public or quasi-public institution would be its best caretaker. The interest in land that seemed to drive these owners was not personal profit but land as part of a human community, values that were often described as "stewardship."

Most public policies seemed to fuel the drive to development, and only a few worked in the direction of land preservation. Some facilitated the transfer of land to a land conservancy by permitting the owner to deduct its value as a charitable gift for income tax purposes; this simply meant that the owners did not realize the full speculative value of the land but a more modest amount. Conservation easements transferred to a land conservancy the right to control the surface development rather than the full ownership, with stipulations agreed to by owner and conservancy as to the limited surface uses that would be permitted. Other policies facilitated such transfers by reducing inheritance taxes for such donations; thus reducing the temptation of heirs to sell land simply to pay the inheritance taxes.

The dominant presence of development pressures on all sides helped to shape the interest in community land stewardship. The interest in nature and natural values had risen steadily in the environmental era, providing a significant basis for instilling them as community values, but the ability to realize those values was limited and frustrating. Nevertheless, the burgeoning interest did lead to a wide range of practical strategies, of which one of the most important was land conservation. That direction of environmental activity was fostered primarily by a sharp reaction to the negative features of development. Although it seemed more than difficult to counter those pressures, it was at least possible to carve out a place in the community for natural lands. And as the private side of land conservation rose, so also did its niche in the realm of public policy.

7 PUBLIC RESOURCES AND PRIVATE RESOURCES

The environment is largely a public affair, consisting as it does of the air, water, and land that people consider to be integral to their daily lives, their homes, their work, and their play. Air and water are more obviously public in that they "belong to everybody," and everybody has a stake in their quality; hence much environmental activity is directed toward the public regulation of how private individuals and corporations use and misuse air and water as a public resource. Land is more mixed in its private or public character; it is "privately owned," which carries certain "rights" along with ownership, but private land use often conflicts with the interests of the surrounding public. Moreover, legal historians argue that private rights to land are derived, both in history and in legal justification, from public approval and supervision of those rights, giving land both a private and public character. This duality leads to a more mixed environmental policy for land, some of which involves the transfer of land from private to public ownership in order to maintain its public value, and some of which involves the public regulation of how private owners use their land in order to ensure that public rights are protected. These issues became more intense as more people vied for use of a limited amount of land.

Throughout the American past the use of privately owned land often gave rise to harmful consequences for other landowners or the general public. Hence innumerable disputes arose as to where to draw the line between the use and the misuse of land, and courts were frequently called upon to sort out those relative rights and wrongs. During the years of rapid economic development in the nineteenth century, courts often argued that the adverse consequences of development were acceptable because of the resulting material public benefits and if people were harmed in the process, that was a penalty society had to pay for material progress. After World War II, however, considerable interest arose concerning the harm that private enterprise brought to the "public," and demands arose for measures to curtail harmful practices. The interest in environmental quality was an integral part of that new attitude.

The use of water and air for waste disposal by owners of private property, such as industry and automobile owners, came under intense criticism amid the desire for a cleaner environment. Wildlife, also a public resource, came under greater protection against those who would destroy it. As a more natural environment came to be prized for home, work, and recreation, those who destroyed forests, wetlands, or open space came to be thought of as destroying a valued resource. In a more nebulous but powerful way, the space around one's privately owned property came to take on a public character with respect to noise, smell, and appearance. In such cases as these, the air, water, and land increasingly came to be thought of as "public," thereby giving people certain "environmental rights." Hence the spate of environmental legislation that took place in the late 1960s and early 1970s represented a historic stage in the public assertion that environmentally harmful use of private property was not acceptable.

Debates over management of public resources range over the gamut of how water, air, wildlife, and land should be used and whether they should be managed by public agencies or private enterprise. Environmental values, which are a part of the new environmental culture, are closely associated with public management for public environmental purposes, and extractive and developmental values are closely associated with private ownership and management and a dominant element of private monetary benefit. In the environmental era, these close connections were intensely debated. Private-market ideological think tanks arose to question the wisdom of public management simply as an undesirable theory; they were closely associated with private developers, extractive industries, and individuals who wished to develop those resources for their private gain. Over the years they presented a continuing challenge to the environmental culture. Yet though they helped to shape public debate, by the end of the twentieth century they had not been able to dislodge the rather deeply felt belief that public environmental objectives could be more fully realized through public resource ownership or regulation.

WILDLIFE

In this complex of issues, we begin with one of the less familiar but highly significant aspects—wildlife. In the United States, wildlife is publicly owned and has been throughout its history. In Europe, wildlife was usually owned by the nobility and royalty, and the general public was excluded from hunting on noble and royal lands and severely punished if caught poaching. To early settlers, one of the more attractive features of land in the New World was the abundance of wildlife. Every settlement, of course, had its privately held lands, but each also included the larger "public lands" beyond, in which game could be hunted freely and which, it came to be felt, individuals had a right to use.

In the nineteenth century, hunting pressures depleted wildlife so much that its disappearance was widely forecast. This state of affairs brought about a sharper definition of the public character of wildlife. Buffalo disappeared early from the East, then in the last third of the nineteenth century declined sharply in the West. Deer and turkey populations, remarked on as abundant in the early centuries, declined in the nineteenth century and by 1900 had almost disappeared from the East. Public game agencies were formed to restore these populations for hunters. At the same time, public revulsion against hunting birds gave rise to protective action to prohibit the use of bird plumage for hats or to provide bird refuges. All this gradually created a perspective that wildlife was a public resource that required public action to sustain, restore, and enhance.

Efforts to maintain and restore game populations were carried out under public auspices, primarily by state governments. One could not hunt any animal at will, but only with a license obtained from the state and only during specified hunting seasons. Furthermore, this license rationed the holder to take only a specified number of that game, and state game wardens were hired to police the system. The state took up responsibility for restoring and maintaining populations by stocking land with deer, turkey, bear, and elk imported from elsewhere. The result of all this was an elaborate system of state game management that was organized and operated through public agencies.

Along with public management came publicly owned habitat for game animals, often called wildlife refuges or public game lands. Some of these came from the desire to establish places where wildlife could live protected from being freely taken. Others were places where wildlife were kept initially to reproduce and then were transferred to "wild" public lands to live as a "wild population" and reproduce naturally. In the case of migratory birds, wildlife refuges served as habitat where they could nest, reproduce, and feed as they migrated from Canada to Central and South America and back again. These activities established wildlife even more firmly as a public resource requiring continuing public management and led to the development of a federal supervising agency, the U.S. Fish and Wildlife Service. Similar efforts were often undertaken by private landowners to establish private hunting grounds, either individually or collectively through hunting clubs. These limited hunting to club members and came to be thought of as overly exclusive. In contrast, publicly owned game lands were thought of as "public hunting lands," open to the wider public and managed in their behalf.

In such incremental ways, wildlife evolved as a public responsibility and a publicly owned resource. Private landowners could post their land against hunting, but they did not own the wildlife so as to be free to "take" game at will. In most states, farmers were given authority to kill wildlife that might damage their crops, even though such authority was difficult to administer. But ownership of land and wildlife were kept separate; ownership of the land did not carry with it ownership of the wildlife that happened to roam or fly onto it. Wildlife in America was considered to be a public resource, owned by the public and subject to public supervision, control, and regulation.

The public nature of wildlife evolved further as the public's interest in wild resources expanded from hunting and game to encompass biodiversity, the entire gamut of plants and animals and their observation, study, and enjoyment. This new approach to wildlife was an outgrowth of an interest in species that had become extinct or were currently endangered or threatened, and it extended to include such resources as plants, neotropical migratory songbirds, coral reefs, wetlands, deserts, old-growth forests, and specific forest species such as lichens, salamanders, frogs, and invertebrates. New programs arose, especially at the state level, to protect species of special concern. Usually called nongame wildlife programs, they often were started with special funding such as income-tax checkoffs or the sale of special license plates. One of the more subtle but significant signs of change was the way in which most state "game" agencies changed their names to "wildlife" agencies.

This broadened interest in wildlife involved a new set of intricate relationships between private and public in attempts to protect a much wider range of resources than simply "game" animals. Were these "public" resources, and were they to be subject to public supervision as fully as game animals had been? Clearly this was the case on publicly owned lands, but did they involve public responsibilities when on private lands? Birds that flew over lands and waters irrespective of who owned them clearly had some sort of public character, but what about those animals that were far more localized in their habitat, such as amphibians and reptiles? And what about plants and plant pollinators? Were they considered important for protection? These kinds of issues pitted a host of scientists and members of the general public interested in the protection of wild resources against the landowners on whose lands they were located.

The federal endangered species programs were especially at issue, because the law required that threatened and endangered species be protected on private as well as public lands. In other words, endangered species were a public responsibility, and the law established a distinction between landownership and wild-resource ownership much like the traditional distinction between private landownership and the public ownership of the game that roamed over it. This broader claim as to the public character of wild resources provoked an intense reaction on the part of private landowners. To those who wished to develop land with permanent structures or to grow crops that might destroy wild resources, there appeared to be little middle ground when it came to the presence of rare and endangered resources on their land. Among many farmers, the reaction was so strong that they refused to permit ecologists to survey their land for rare and endangered species on the grounds that it would surely lead to restrictions on what they could do with their land.

PUBLIC WATERS

Public waters—rivers, lakes, ocean shore waters, wetlands—also had a long-standing tradition of public ownership in which private use was permissible but only under public regulation and supervision. Some of the firmest legal statements of public responsibility came in regards to shore waters, which, in English common law, had been a "public trust" owned by the Crown on behalf of the public, rather than "alienated" to private ownership. That tradition had continued in America and was the subject of a rather famous court decision in which Chicago could not "alienate" Lake Michigan lands under waters near the shore to fill them and construct a railroad on the grounds that governments could not give away their responsibilities for such lands. They were a "public trust."

The most traditional use of public waters, however, was for fishing. The earliest settlers found the rivers filled with fish, which served as a significant source of food equal to that served by game animals. However, as was the case with game, over the years, rivers were overfished or polluted and those concerned with a continuous supply of fish sought to restock waters with young fish grown in fish hatcheries. But in this case, the fish managers did not enjoy the same possibility as game managers did when stocking such game as deer and turkey that it would soon lead to natural reproduction. Instead, they adopted the practice of continually replenishing stocks from hatcheries. Only in a few states did the "natural trout stream" become an important element in a diversified management program in which the higher-quality trout streams were not stocked and provided opportunities for trout fishing in a more natural setting.

Over time, use of rivers and lakes for fishing was continually compromised by the constant encroachment on natural waters by commercial enterprises. Rivers and lakes had long been used by logging companies as free transport and temporary storage for logs. Small-scale mills for grinding grain and sawing logs used falling water for power, and as the new factories of the nineteenth century sought to use water power, more rivers were dammed to increase the availability of falling water. Private utilities proceeded to dam rivers to generate hydroelectric power, and in the twentieth century massive federally funded projects dammed the main stems of rivers for multiple uses combining navigation, flood control, irrigation, hydropower, and flatwater recreation. After World War II, the U.S. Soil Conservation Service added to this use of rivers by damming smaller headwater streams for multiple uses including wetland drainage as well as flood control and reservoir-based recreation.

One of the most significant elements of the enhanced environmental culture of the last half of the twentieth century, however, was the notion that rivers had many uses other than development. Some of this involved an expanded interest in fishing along with the dawning idea that dams and encroachments had undesirably disrupted the migration of fish and that building such structures should end in order to restore those migration patterns. Increasingly, rivers were used for water-based recreation such as canoeing, kayaking, and river rafting. These activities led to federal and state programs to identify and manage particular segments of rivers as "wild and scenic" rivers in which the natural values of space as well as river recreation became important objectives of river management. And still other objectives involved an interest in water-based endangered species, such as fish and mussels, which a more natural river could help to restore.

Many changes in the use of waters came about through both public and private means as environmental values shaped new policies. Some of the more striking of these involved the old logging ponds of the northeast, many of them glacial lakes and ponds and others enlarged with dams for commercial purposes. In the years after World War II, the growth of state parks, which had originally emphasized scenic land resources, came to include water-based amenities. Many a lake formerly used for commercial purposes now became a public park, with associated efforts to end its commercial use in favor of these amenities. Many factories were located on rivers and lakes and used public waters as a waste-disposal site without charge. This brought to public waters a mixed private and public character in which private parties used a public resource for their commercial benefit without payment.

Much of the transition in public waters came about as private individuals purchased land along lakes and rivers for their amenities, either as permanent residences or as vacation sites. Some of this involved water-based recreation, but much also involved the passive environmental values of a more natural living space with some respite from the congestion of urban living. Often those who lived around lakes or in a river watershed found a shared interest in the resource they all valued in common. This led to the formation of lake and watershed associations with common goals to monitor water quality and protect the natural values of their water-based environment. Frequently an incremental transition took place from the old logging camps to newer tourist and then vacation home sites, which sharply transformed waters from commercial to scenic and recreational resources.

Amid these private changes, however, the public legal status of rivers and streams remained. Those who sought to use waters for their own private purposes found that the public implications of waters presented limits to their activities. If they wanted to modify a wetland along a river or lake, they found that they could do so only under public restrictions. If they wished to restrict access by the public to the use of waters for fishing or recreation, they could not fully do so. The environmental era, in fact, gave rise to a persistent, even though slow, revival of the traditional public interest in waters—a revival that was not always clear because now a host of new public uses that involved varied environmental amenities often competed with each other. One of the most common conflicts was between those who canoed and those who fished, who got into each other's way as the intensity of use increased. These competing private uses now had to be worked out within a combined private and public use of a limited public resource.

Western rivers provided a marked variation on this pattern. There the scarcity of water led to the development of new relationships between public and private rights. In the earliest years, few public legal rights to waters were firmly established, so those who wished to use waters simply diverted them from streams for their own use. Toward the end of the nineteenth century, states, starting with Wyoming, began to regularize this process by establishing a public water agency that sorted out private rights within a context of ultimate public ownership. The most significant doctrine was "prior appropriation," a process whereby those who filed claims to water first had prior rights—first come, first served—with some stipulations that this would hold only so long as the water was "beneficially used." Slowly this procedure was worked out within the context of knowledge about how much water was available, how much was not yet appropriated, and what uses were considered to be "beneficial." Water rights were appropriated by irrigators, by industries, and by municipalities, and their relative claims were adjudicated within the context of public supervision.

These traditional forms of western water use and rights were subject to the same pressures from new environmental uses as was the case in other parts of the country: fishing and river rafting required water, usually thought of as "in-stream uses" in contrast with the extraction of water from rivers for use by land-based activities. They were also subject to increased demand for water from cities, which grew rather rapidly after 1970, causing intense controversies between older agricultural and extractive users and new urban users. Often the political demands of cities came to express both the new consumptive uses of urban dwellers and their nonconsumptive interests in river recreation and amenities. Much of the controversy revolved around what was a "beneficial use." Were in-stream uses for fishing, river rafting, or amenities "beneficial uses"?

Many of these new uses as well as the continuing public status of waters came into sharper definition in the 1980s and 1990s as a host of previously licensed dams, most in smaller streams, came up for relicensing. Former decisions had revolved primarily around the dams as sources of hydroelectric power. But now, as a result of new legislation in the 1980s, a number of environmental considerations had to be taken into account by the Federal Energy Regulatory Administration (FERA) in decisions to relicense. Commercial firms and municipalities that owned the dams sought to restrict sharply the role of competing uses, but the courts upheld the legislation that required FERA to take into account a broader range of interests in making its decisions. A major consequence of this new setting for relicensing was that a wider range of local interests was activated into participating in relicensing decisions, shaping quite a new political context for the relevant decision making. In some cases, this led to proposals that the dams should be eliminated, a proposal that was appealing when compared with the extensive cost of reconstruction and renovation.

As the endangered species program came to be applied to fish, the West, particularly the Pacific Northwest, faced a distinctive new set of river-related values, intensely debated within the context of public water resources. At issue was the salmon as an endangered species and the adverse effects of Columbia River dams on their migration. A wide spectrum of people in the Northwest—Native Americans, participants in the fishing industry, sport-fishing enthusiasts, and environmentalists interested in more free-flowing rivers—came to focus sharply on the problem. Save for a few cases of headwater dams, few proposals aimed to eliminate the dams, but there were many proposals for new management schemes to facilitate fish migration and fish spawning, all of which meant that use of dams for hydroelectric power would have to be modified in order to accommodate the fish. At the same time, during the Clinton administration, forest policies in the Pacific Northwest were shifted somewhat toward watershed protection as a device for helping to protect the fishery.

PUBLIC LANDS

For many years the federal public lands, primarily in the West, have been a major focus of environmental politics. Most had been acquired through various past treaties with Indians or with other countries, some after the Mexican War, or when states ceded their lands to the federal government. Early national policy was to sell this land to private individuals to encourage settlement and resource development. As time went on, however, sentiment arose to cease private disposition and to support permanent public ownership and management. Lands were "reserved" from sale, as the phrase went, first to form national parks, then national forests, then wildlife refuges, and finally in the mid-twentieth century, the remaining lands—by then called the "public domain"—were withdrawn from sale to undergo permanent management primarily as grazing lands. The last debate over this issue involved the vast acreage in Alaska, but it ended with the Alaskan National Lands Act of 1969, which distributed the acreage there to Native Americans, to the state of Alaska, and to federal land agencies.

The environmental era brought a new phase to the history of federal management of these lands. Prior to World War II, resource extraction and development had dominated federal management, especially of national forests and grazing lands. Quite different attitudes had arisen around the formation of both the national parks, in which preservation and use for human enjoyment had dominated policy, and the national wildlife refuges, which had been thought of primarily as habitat for wildlife. With the rapid growth of outdoor recreation, including fishing and hunting, hiking and camping, the national forests began to include a marked recreational component that was initially recognized for the national forests in the Multiple Use Act of 1960. At the same time, a movement arose to reserve some national forest lands as wilderness areas, free from human development and permanent intrusion, a move that was realized in the Wilderness Act of 1964. Wilderness areas were also authorized on the other federal land systems. As time passed, an increasing amount of land was incorporated into the National Wilderness Preservation System.

The long debate over whether or not federal lands should be added to by public acquisition of private lands instituted a new phase in public land management. This was especially the case in the East, where little publicly owned land remained and where public management would require significant acquisition. At the start of this venture in the early twentieth century, the strategy did not involve significant tension, since many private lands in the East were in the process of being abandoned because they were unproductive for farming or because their timber had been cut and they were no longer wanted by timber companies or because they were held by timber-using industries such as the charcoal iron industry, which was rapidly becoming obsolete. The first federal purchases of such land began under the Weeks Act of 1911, and in the ensuing years, some 25 million acres of formerly forested land were acquired by federal purchase for management as national forests.

This was only the largest of a variety of acquisition programs. Some involved the acquisition of wetlands for migratory game birds, a program that was strongly supported by hunters and was funded by a federal duck stamp required from each hunter. Wildlife refuges for game birds were established up and down the United States along the routes of four major flyways that extended from Canada to Central and South America. Others involved state acquisition of forest lands, especially in New York, Pennsylvania, Michigan, Wisconsin, and Minnesota. The largest block of these came from farm and forest lands that had been abandoned by owners because of their marginal productivity, land that then was forfeited and reverted to state ownership to become state forests or, in the case of Wisconsin, to county ownership. Still other acquisitions were made by the National Park Service through its program to establish National Seashores and Lakeshores, a program that extended the traditional park program for land areas to significant coastal recreation and natural areas.

The public applied its growing interest in nature to each of the land systems and sought to inject nature protection into their management. In each case, this led to challenges to extractive and commodity uses that had been customary on many such lands as environmental and ecological objectives now called for their curtailment. Ecological science had grown steadily over the years, leading to a wide range of specialists who produced new knowledge about the composition and functioning of natural systems. These professionals came together in various scientific bodies to support management for natural objectives, and this in turn called for shifts from a dominant commodity/extractive focus to a greater play for natural processes. These shifts were more readily accommodated on the national park lands and fish and wildlife refuges, and were far more difficult to bring about on national forest and grazing lands, where extractive and commodity activities were more firmly established.

Controversies over public land management often took the form of conflicts between federal and state authority in which one of the major arguments was that federal lands should be transferred to the states. Some of this arose from the belief of mining, grazing, and timber interests that they would fare better if the lands were managed by the states. On the other hand, in western states where there were large acreages of both state-owned as well as federal land, recreational and environmental interests preferred federal policies to existing state policies and helped to prevent measures to transfer federal lands to the states.

As environmental and ecological values came to influence public land management more fully, the federal government took steps to recognize the needs of species whose life histories required them to move not only between states but between nations. This led to federal migratory bird treaties, the first with Canada in 1917, and also to federal action to establish national wildlife refuges along the north-south flyways. At the same time interest arose in expanding wildlife programs to include "nongame" wildlife. State agencies moved ahead with these programs sufficiently to call for federal programs to supplement and assist them financially. Two federal programs raised more substantive issues in federal-state relations. One was the endangered species program, on which states had moved forward slowly, but which now proceeded with strong federal initiatives; some states complemented federal endangered species programs with their own programs to protect species endangered within their states. The other involved management of wild species on federal lands. By law, wildlife belonged to the states, but tension arose when federal and state management policies differed, as when federal agencies sought to exert authority to protect wildlife on their lands from what they considered to be state mismanagement. State agencies, in turn, denied that the wildlife-habitat connection called for separate federal initiatives.

More fundamental and underlying issues in public land management involved tension between private market objectives and public environmental and ecological objectives. These usually drove the controversy between federal and state authority and displayed a more fundamental life of their own. The private business community spearheaded the conversion of natural resources to developed resources, and the environmental community believed that that influence would have to be curbed in some fashion if the protection and preservation of nature were to advance. This promoted the continual belief that only if land were owned and managed by the public would more natural values prevail.

PRIVATE MARKETS AND PUBLIC RESOURCES

The attempt to recognize the public nature of environmental resources faced continual opposition from advocates of private enterprise. The most dramatic were the theorists who argued simply that resources of all kinds, environmental and material, should be left to private enterprise alone. Private ownership and use of resources provided the highest benefit for the public; public ownership or regulation was not only undesirable but downright dangerous. Free-market theorists argued that private landowners, on their own initiative, provided sufficient environmental benefits for both present and future generations, and that public ownership would be a burden rather than a boon. Publicly owned lands, they argued, should be sold to private owners, and environmental regulations should be abolished. As the call for public action to foster environmental objectives rose after World War II, so also did these assertions of free-market theorists.

In close association with private-market theorists were property owners who had direct economic interests at stake and who argued against governmental ownership or regulations. They did so on a selective basis, for they were just as supportive of governmental action that would benefit them as they were against governmental action that would restrain them. Some simply opposed regulations such as wetlands or endangered-species protection on private lands on the grounds that they were public infringements on private property, and they joined with the free-market theorists in refusing to recognize the intermixed private and public character of environmental air, water, land, and wildlife. Miners insisted on long-held rights to acquire and mine on federal lands for a pittance; western cattlemen argued that because they had grazed public lands for years, they owned them and could therefore control their management. Still others sought to benefit from the natural amenities provided by public lands by developing lands on their borders to take advantage of the added monetary value provided by the proximity of public lands to their private holdings.

It was often difficult for those in the private market to accept the notion that public natural resources were an important public asset. In the years prior to World War II, the timber industry supported expansion of the national forests as well as a policy of not using them for timber production because they wished to separate these lands from competition with their own private resources. After the war, when they had depleted their own wood supplies, they began to look to the national forests to augment their own reduced supplies. As the public began to turn to environmental and ecological values as the major benefit of forests and to consider wood production to be a minor though important use, the timber industry objected strenuously to the resulting implications for forest management. At times they sought national legislation to require a given level of continuous wood production on the forests; at the state level they sought programs to reduce taxes on private holdings, to increase production from public lands, or to use public funds to enhance the industry.

In a number of cases, the intermixture of public and private resources created dramatic controversies. One was the presence of inholdings, or tracts of land under private ownership, in the national parks. Over the course of the history of national parks and forests private inholdings had persisted, and there was a continual push to facilitate public acquisition of these lands to improve public management. One problem was the constant pressure from private inholders for road access over public lands to their holdings. From time to time inholders were quite willing to sell, but many were not and continued to object to efforts to encourage them to sell. The new national lakeshores and seashores presented particularly knotty situations, because most of these combined holdings of existing federal and state forest and park lands with acquisition of private lands, and over the long run this involved pressures on private owners to sell. The terms often permitted private ownership to continue for several generations, with ultimate sale to the parks to occur in the future. And in some cases, although owners had agreed to such long-range terms, as the eventual year of ultimate sale approached, they attempted to reverse their initial agreements.

Assertions as to the desirability of the private market in land and other resources continued to be made throughout the environmental era. On the whole, however, these arguments did not meet with sustained success largely because the actual practice of resource use by those in the private market was often unacceptable to the public. Actions of private landowners seemed to reinforce the view that only through public ownership and management could public environmental objectives be met. The continued argument by free-market advocates that private owners used resources without harm to other people or resource values and that they were quite able to preserve resources for future generations seemed to fall on deaf ears, especially when the public regularly learned of cases to the contrary. There was sufficient evidence on all sides that those in the private market were not reliable stewards of wildlife, waters, and land, most of which, if subject to market forces, would be dissipated through pressures for short-run private gain.

In these issues the primary focus was on land, because environmental values seemed always to come back to a given place where wildlife, water, or land resources were located. Public land is central to public environmental values. Over the years, therefore, the commitment to public lands did not abate; instead, in many states, the drive to increase the conversion of private land to public land advanced steadily. A host of devices arose to find the funds to bring about such acquisition. These began at the federal level in 1964, when Congress established the Land and Water Conservation Fund, derived from proceeds from oil and gas drilling in offshore waters under federal jurisdiction. From this fund monies were disbursed to federal, state, and through state action to local governments to acquire land. Under budgetary constraints after the early 1980s, Congress steadily reduced this commitment and diverted offshore oil revenues to balance the federal budget instead. In response, demands arose at the state level for special funding programs to acquire land. A variety of sources were tapped: state lotteries; proceeds from extraction of state-owned mineral, oil, and gas holdings; taxes on real estate transactions; and bond issues. By the 1990s such strategies had become more widespread at the county level as well, as communities felt that if left to private hands, all open space in their community would be destroyed by development.

In the 1970s and thereafter, a new strategy, the land conservancy, evolved as a device to protect land from development pressures of the private market. Land trusts were organized under private auspices to acquire land for permanent preservation and hold it free from private-market forces. In some cases land trusts acquired land by gift; in others, by purchase or combined gift and purchase. The process was facilitated by a significant exemption from federal taxes, which with time became more elaborate. The value of gifts of land or sale below market prices could be deducted from federal income taxes and later also from federal estate taxes. This greatly stimulated land trust acquisitions. At the same time, local governments contributed to the process by exempting land trusts from property taxes on the grounds that they served important public functions. In almost all cases one of these functions was educational, as the land trusts developed extensive programs in nature education for schoolchildren as well as adults.

Land trusts were first established in the 1970s but by the end of the century grew to over eleven hundred. They were started first in New England and the Middle Atlantic states and then extended their range throughout the nation. In many cases state funds for land acquisition were granted to private land conservancies as well as to public agencies, thereby augmenting the conservancy's role in land conservation. Equally important were the funds provided by private foundations, established by people of considerable wealth who had a deep commitment to land preservation, a collective activity known as "conservation philanthropy."

The land trust identified more sharply the strong attachment of the public to nature protection by building extensive memberships and collecting financial contributions from a wide range of people. That attachment was reflected also in the large number of people who sought to transfer their land to a land trust and who in the process frequently gave up the opportunity to profit fully from private-market sale.

Land trusts also sharpened the resistance to public ownership of land on the part of those with theoretical commitments to the private market. They found the private ownership of quasi-public land conservancies to be as objectionable as public ownership itself. By the 1990s organizations hostile to public landownership had focused their attacks especially on the Nature Conservancy, the national organization with the largest land conservation program. Such groups succeeded in reducing federal funds for land purchase, but by the end of the century they had not yet succeeded in slowing down state purchase programs and had made no dent at all in the continued popularity of the land conservation trust. They frequently argued that state land holdings were sufficient and should not be augmented, or supported a "no net gain" policy in public lands so that when more were added, others would be sold. But amid such arguments the public assumption that public lands were needed for many public objectives continued strong.

One could well chart an increasing tension between public and private ownership and management in wildlife, water, and land over the years and predict that such tension would increase. The underlying circumstances of this tension were simply the increasing demands on finite resources arising from a growing population with growing levels of consumption. While land and water as basic resources and the wildlife habitats they provided remained constant, the number of people using these resources continually increased. Amid this fundamental dilemma a major change took place from a predominant emphasis on resource extraction and development to a growing interest in resources for appreciative and learning purposes. Hence the public continued to view the private market as limited in its ability to provide environmental benefits and continued to rely on public resources for that purpose.

8 ENVIRONMENTAL ENGAGEMENT

When one thinks of citizen involvement in environmental affairs, the term "environmental movement" often comes to mind. Usually this comprises the citizen environmental organizations and their activities, some general ideologies that are used to justify and advance attention to environmental affairs, and the public policies that result from political action. The term "movement" has been applied to similar types of social and political action in the past, such as the "labor movement," the "prohibition movement," the "civil rights movement," and the "women's movement." The customary set of "movement characteristics," as mentioned above, is attributed to the "environmental movement."

For environmental issues, however, such a set of movement ideas is more than inadequate. Environmental action is a type of "movement" action in that it seeks to arouse and increase public support for public policies and uses the customary types of education and organization to do so. But environmental activities are far more extensive than social and political action and are more inclined to involve concrete activities pertaining to environmental resource conditions rather than just ideological argument and commitment. Formal ideas play a relatively minor role in environmental affairs and are far more overshadowed by concrete environmental circumstances. Hence the term "environmental engagement" is far more appropriate than "environmental movement."

The larger dimensions of environmental affairs are best described as an "environmental culture" rather than an "environmental movement." They go far beyond the formal organizations usually taken to represent the "movement," to values and ideas that have infused many social, economic, and political institutions. They involve meaning, how people think, and what people value, ranging from vaguely felt beliefs and commitments to far more intense ones. They imbue education and religion, science and law, some economic ventures, a host of professional activities, and individual action all the way from recycling to a wider range of learning and varying degrees of political action. Though some of this includes involvement in "movement" activities, much does not; rather it entails a wide range of personal activities and institutional developments in quieter and less obvious ways.

ENVIRONMENTAL ORGANIZATION

We begin first with the environmental organizations and then go to the broader context of environmental engagement. When most observers think about the environmental movement, they think primarily about the national organizations headquartered in Washington, D.C. These have been referred to often, first as the "big six," then the "big ten," and more recently, as their number has increased, to the "twenty-six" of the "Green Lobby." Generalizations about the movement are based largely on these groups, their membership size, their policies and programs, their successes and failures. The resulting conclusions, however, are faulty, because environmental organization is far more extensive, runs from local to national levels in close networking fashion, and tends to proliferate into specialized groups that take on only one or a limited number of issues. By and large this network of interacting organizations is poorly understood because of the almost universal focus on the organizations headquartered within the Washington, D.C., "beltway."

State-level environmental organizations deserve particular attention. They were formed as early as the national organizations, often separately even though in cooperation with them. Examples include the Washington, Oregon, and Idaho Environmental Councils formed in the late 1960s; the Maine and Vermont Natural Resource Councils formed about the same time; and Florida Defenders of the Environment, which took on the task of transforming the Cross Florida Barge Canal into a regional environmental asset. Some of these received organizational help from national organizations, such as the state councils in the Pacific Northwest that were encouraged by the Sierra Club or the Maine and Vermont Councils that were state branches of the National Wildlife Federation. In the late 1970s and early 1980s, a further spate of state organizations grew up in the Midwest, the South, and the East, many of them to focus specifically on the emerging environmental issues within their own states and especially to engage in lobbying activities in state legislatures.

The drive in all this organization came from the environmental activities within the relevant states. State issues, state energies, and state financing impelled both the extent and the character of the state organizations. Consequently, the organizations varied enormously in their organizational strength and took on quite different issues depending on the course of events within each state. The result was a landscape of environmental organization that varied from region to region. States with a stronger environmental culture gave rise to a much more vibrant environmental organization, while those with a weaker environmental culture were less developed. Among the first were the states of New England, including both New York and New Jersey; the Great Lakes; Florida; and the Pacific Coast. Among the second were the states of the Gulf Coast and the Great Plains.

These state organizations played a significant role in bringing together local and regional groups within the states into more cohesive relationships. Often the state organizations took the form of coalitions, which were shaped by representatives of the groups that formed them, gave them financial support, and formulated their policies. The Sierra Club, which was the main national organization with a comprehensive set of state chapters, played a significant role in developing more effective statewide coalitions to act in state legislative politics, serving either as members of such coalitions in some states or providing the main focal point of political action in others. As the national organizations grew in membership and resources in the late 1970s and the 1980s, they established regional offices that promoted regionwide issues and brought state organizations together. Often they were crucial in giving state groups more technical resources and enabling them to cope with national issues relevant to their regions.

Over the years, therefore, a progressively more elaborate set of organizational activities arose throughout the nation with a fairly well developed vertical structure. At the state level, local, and state, and regional organizations networked for statewide action; at the national level, state and national organizations networked for national policies. A wide range of multilevel organizational impulses led to a scramble for scarce resources at each level. Each organization sought to increase its capabilities, and this meant a constant struggle over limited resources. This elaborate pattern of organization involving many varied and diverse independent groups, each with its own clientele and sources of support, was accompanied by shared views that they were all part of the same effort.

Environmental organizations were continually challenged by their membership to do more. On the one hand, the range and intensity of citizen interest in environmental affairs continued to rise, and the organizational and technical tools demanded by members increased. Yet the organizations, with finite resources, found that they had to be selective in their activities and could respond only in limited fashion to the higher level of environmental engagement that citizens wanted. The organizations came under continual fire from members who wished resources to be diverted to their issues. This tension was not felt as much by the single-issue organizations such as the National Parks and Conservation Association, the Defenders of Wildlife, the Nature Conservancy, or the Wilderness Society, but was felt more by the multi-issue organizations such as the Sierra Club, the National Wildlife Federation, the Audubon Society, the Natural Resources Defense Council, or the Environmental Defense Fund. The tension gave rise to considerable internal debate and struggle over how limited resources should be spent. In the case of the Sierra Club, which had the most elaborate vertical organization and invited wide internal expression of opinion, it led to the requirement for annual membership-wide priority setting.

These internal tensions continually actuated one of the main features of environmental organization: the establishment of new and specialized groups. There were specialized groups dealing with coastal pollution, pesticides, indoor air pollution, farmland preservation, mining, population, or climate change. Many of these were formed through the initiative or support from a larger multi-issue organization that desired some sort of action but knew their own resources would not permit it. Specialized groups also arose to deal with a particular kind of wildlife, such as bears, wolves, raptors, frogs, whooping cranes, bald eagles, bats, butterflies, lichens, invertebrates, and elk. These combined scientists interested in research on each species and citizens interested in advancing their own biological knowledge and in providing financial support for species-specific promotional and scientific work.

Most observers do not take into account this larger organizational scene largely because they are not acquainted with it. It is far too extensive and complex for them to cope with, and they do not follow its activities through the many available specialized publications. Many academic observers, most media, and professional environmental watchers do not spend the time and energy needed to observe environmental organization carefully; thus they collectively miss the mark and arrive at what are quite ill-informed conclusions.

Environmental organization had both regional and topical variations, and these became more elaborate over the years. The regional variations were especially significant. In the Chesapeake Bay area, for example, environmental culture was organized around the protection of the Bay; in New York, around the Adirondacks; in California, around the Sierra Nevadas; and in the South, around wetlands and swamplands. In the Great Plains, natural environment issues were far less salient, and the agricultural and resource-extraction base of the region gave rise to the term "natural resource council" rather than "environmental council" in order to strike a more responsive chord from people in a region with a more limited environmental culture.

Environmental organization also became more elaborate over the years, expanding energies and capabilities into a greater variety of issues. By the year 2000 the environmental network was far more complex than it had been four decades before. The year 1994 witnessed an important step in this elaboration, as the Republican Party attempted to modify sharply much of the environmental program of the previous three decades and confronted the entire environmental community with a hostile legislative climate. Few environmental organizations were prepared to add this dimension of action to their repertoire, and because most were charitable organizations under the rules of the Internal Revenue Service, they were free to engage in some legislative and administrative lobbying, but they were not free to engage in electoral politics. Two exceptions were the Sierra Club, which had given up its charitable status in 1968 precisely to be free to participate in a full range of political activities, and the League of Conservation Voters, which since 1970 had produced legislative voting charts as an aid to voters who wished to vote for candidates who had demonstrated support for environmental objectives. In 1995 a new citizen-based group, Republicans for Environmental Protection, was formed to focus especially on raising the level of support for environmental policies among Republicans in Congress.

THE ENGAGEMENT OF ENVIRONMENTAL KNOWLEDGE

Underlying environmental organization was a more pervasive set of environmental engagements that ranged from the popular and professional search for pertinent scientific knowledge; to the expression of environmental aesthetics through art, photography, and literature; to the environmental implications of religion; to the search for knowledge about the way in which public decisions are made. Other specialized environmental activities helped to build up a varied environmental culture that included the environmental design of living arrangements, the development of wilderness medicine, mystery novels with an environmental setting, and the establishment of environmental media beyond the publications of environmental organizations. Together, this wide range of activities added up to a general extension of thought, meaning, personal expression, and use of the mind to explore and express environmental ideas.

An interest in acquiring knowledge and thinking about environmental issues was one of the most compelling ingredients of environmental culture. A host of institutions and activities were involved in shaping it: the systematic research conducted by professionals; the search for knowledge about the environmental issues in which one was engaged; the role of the media in shaping what one thought about environmental affairs; nature education in the schools and nature centers; environmental monitoring organized by citizen groups; the range of available books and pamphlets; individual environmental research carried on through personal computers; and knowledge of the political process as issues moved through various stages in the course of decision making. Environmental knowledge—the search for it and the desire to make it relevant—was one of the overarching features of environmental culture, a process in which people from all walks and areas of life tended to participate with the excitement and satisfaction of discovery.

We are often inclined to think of this as a process in which scientists discover knowledge, then seek to educate people about it, and then attempt to apply it in various ways. But the search for environmental knowledge preoccupies many different people and tends to establish a common ground of interest and engagement. This does not imply a firm agreement on what that knowledge is, but only a common interest in and a common pursuit of knowing. In some cases, it involves young people's choice of a career in which they can apply environmental knowledge and skills; in other cases, it involves choice as to avocation, or the desire to know more about the piece of the natural world in which one is particularly interested, such as birds or wildflowers; in other cases, it involves an attempt to understand more fully the environmental circumstances of public issues; and in still other cases, it involves attempts to educate children and young people about nature and how it works. All this constitutes a widespread desire on the part of a large number of people to engage with their world by learning more about it.

In this search for knowledge, environmental education for the young was one of the most compelling features, reflected in school environmental education, the development of nature centers that also provided environmental education for schoolchildren, and the desire of state and federal natural resource management agencies to adopt and foster environmental education programs. As people grew older, the search for knowledge often took on a more specialized focus, as it was impossible to became knowledgeable across the board. Most attractive was to become more knowledgeable about a particular group of animals or plants; there were amateur botanists, specialists in bears and wolves, salamanders and butterflies, lichens and frogs. This specialized interest generated organized efforts to integrate one's search for knowledge with action to foster scientific research on the species and to support protection programs. Some came to such activities through hunting, such as the Elk Foundation; others came to it through appreciative watching, such as observing neotropical migratory songbirds. Often environmental education for the young evolved naturally into the avocational search for knowledge as one grew older.

Citizen environmental monitoring was closely related to these activities and brought different ages together. Initially citizen monitoring focused on water quality. In environmental education classes in primary and secondary education and also in higher education, water-quality monitoring came to be a major strategy for exposing young people to the scientific problems of understanding water quality and, at the same time, learning the techniques for measuring it. The process gave young people a positive feeling for nature and fulfilled their desire to know more intellectually about the way nature worked. But monitoring did not stop with formal education; it often also became a part of adult lives as individuals of all ages learned how to take samples under the guidance of scientists, then adopted a stream or a lake to sample continuously, using certified testing laboratories to determine the results.

Citizen water-quality monitoring took place in streams, lakes, bays, and coastal waters, and networks of such activity developed among school groups across states and countries. Some who began with an interest in stream chemistry extended their observations to stream fauna such as frogs and other amphibians; some located and marked changes in wetlands, both in their water quality and their plant and animal life. Amateur botanists found excitement in monitoring forest conditions, such as spring wildflowers, the annual growth of trees, or progress in forest regeneration after timber cutting. Often citizen monitoring was limited to those measurements that could be carried out inexpensively, since many monitoring instruments were quite expensive. Yet some citizen groups acquired instruments to monitor emissions around nuclear power plants or to monitor ground-level ozone.

The more satisfying stages in citizen monitoring came when it was used as a basis for programs of environmental improvement. Environmental management agencies were suspicious of information acquired by citizens. For this reason, citizen monitoring activities invariably were organized under the direction of a professional specialist and carried out with their guidance. The Environmental Protection Agency, recognizing that water-quality data needed for effective programs was limited, gave considerable encouragement to citizen monitoring; states that implemented water-quality programs were slow to take advantage of this source of labor, but when some did, it greatly augmented their management capabilities. The underlying problem was that though every water-quality agency had a minimum ability to identify major water-pollution sources, few were able to sample tributaries, which often were the sources of main-stem river water-quality problems. Hence citizen monitoring efforts filled a massive gap, providing volunteer labor for tasks that might otherwise not be taken up because of personnel costs.

As these citizen efforts in environmental monitoring progressed, one long-term set of citizen monitoring data emerged as a gold mine of information: the annual bird counts conducted under the auspices of the Audubon Society. These counts began in the 1890s and by the latter part of the twentieth century had provided an extensive set of long-term data. This data remained relatively obscure for many years on the grounds that it was collected by amateurs and therefore was not reliable. However, after World War II, scientists began to analyze the information. From this work came the first conclusions in the 1980s about changes in the populations of neotropical migratory songbirds.

Much of the knowledge component of the environmental culture was organized around specific environmental issues and circumstances that those engaged in environmental activity sought to learn more about. A community or group of individuals would be confronted with a specific issue, such as water or air pollution or development in a wetland. These individuals then endeavored to learn more about the causes of the environmental degradation. Citizens read scientific journals or enlisted the help of professional specialists in academic institutions. Their knowledge was frequently shared with others, assisting in creating a culture of environmental knowledge. As the computer age advanced, some took up the challenge of organizing available data to understand a set of environmental circumstances in ways not understood before.

This growing knowledge involved not only environmental circumstances but also knowledge about environmental decision making, the ways in which the political process worked in the ebb and flow, the give-and-take, of public debate and action. This was a subject in which the citizen environmental organizations, both national and state, were quite active. For, though the general public was interested primarily in more general levels of knowledge and action, the environmental organizations themselves were deeply involved in the details of policy and political action. The knowledge they contributed to this phase of environmental affairs by following national, state, and local issues was unique.

Some of the most significant focal points of environmental culture were efforts organized around specific regional features such as a bay, lake, or river, a mountain range or a barren, a wetland or an area of distinctive plant and animal life. The Chesapeake Bay, for example, spawned the development of a "Bay consciousness" shared by many around the Bay. It involved extensive educational programs in which young people were given instruction in water-quality science and technology through regular classes on boats that toured the Bay. The Bay program focused on submerged aquatic vegetation as an indicator of water quality, so human attention was continually engaged in following the ups and downs of these plants to reveal progress in water quality. The program also broadened attention from the water of the Bay itself to the role of land-use practices on forests and farms around the Bay and the contribution to Bay water quality from tributary streams in a watershed-wide perspective.

The media made their own contributions to the development of the environmental knowledge culture. In the years after World War II, the publications of the national environmental organizations were almost alone in writing about environmental subjects, and they dealt primarily with information about national issues. The local news media provided some of the most extensive information about local issues. The national news media, on the other hand, contributed far less to the fund of substantive environmental knowledge because of their penchant to organize information around opinion and debate rather than to provide details essential to understanding the topics. Much of their coverage was infused with the details of debated environmental ideology rather than with substantive environmental circumstances. The magazines of national environmental organizations continued to be the main sources of knowledge about national affairs, as were the publications of state environmental organizations crucial to understanding state environmental affairs.

With time, some new media ventures, frequently organized around regional affairs, began to make significant contributions to the culture of environmental knowledge. The oldest and perhaps best known of these was *High Country News,* a newspaper independent of environmental organizational affiliation and free from the burdens of advertising income, that provided unique coverage of environmental issues in the Rocky Mountain region. By the end of the 1990s one spin-off from *High Country News* had occurred with the *Cascadia Times,* published in Seattle, Washington. A weekly of more general circulation, the *Maine Times* also was distinctive in its degree of environmental coverage. In a more focused way, the *Bay Journal* (Chesapeake Bay), which combined coverage of scientific and technical information with policy and program information for the Chesapeake Bay, served as one of the best examples of environmental knowledge organized around a specific regional set of circumstances.

POLITICAL ENGAGEMENT

Knowledge was certainly one feature of environmental engagement; many, however, sought to combine understanding with action leading to environmental protection or improvement. Although many environmental circumstances could remain personal, most involved the community, the state, the region, the nation, or the Earth as the environment within which such protection or improvement would take place. Hence the world of public affairs, which was encompassed by the reality of environmental politics, attracted many, and political engagement became a major feature of environmental affairs.

Environmental political engagement was based less on ideologies and more on practical circumstances. Intellectuals concerned with environmental affairs were prone to think in terms of ideologies, and, as was customary with many past political movements, this gave rise to intense debate over relevant ideas. But in contrast with past labor-based movements, environmental intellectuals produced few journals devoted to the presentation and debate of environmental ideas. *Environmental Ethics,* sponsored by philosophers, comes to mind as one of the few cases. Far more were devoted to a range of "how to do it" issues, expressed in the form of books and magazines fostered by environmental organizations. *Island Press,* one of the few distinctive environmental publishing ventures, was devoted exclusively to books intended to give readers ideas about how to protect and improve the environment rather than how to think correctly about it; it rightly advertised itself as the "nation's foremost environmental publisher."

Political action involved a wide range of activities: efforts to ferret out information about issues, guides to practical action, reports on the conduct of legislators and administrative agencies, legal action, dissemination of scientific knowledge—all of which were geared to more effective environmental action. Of all of these perhaps the most widely developed and followed was environmental lobbying before Congress, state legislatures, and administrative agencies. Almost every environmental organization found that legislative and administrative decisions were vital to their goals. At first, environmental lobbying activities were sporadic, arising from the interest of citizen groups in particular legislation relating to air quality, water quality, wilderness, forest or grazing land management. Soon these occasional activities developed into more continuous legislative action with paid staffs.

Few legislators could know much about issues save those in which they specialized, and those who did were relied on by their fellow legislators as experts. Since legislators and their staffs had only limited time and resources, information brought to them by lobbyists often provided the crucial basis for policymaking. Lobbyists, in turn, were influential only to the degree to which the information they provided was reliable and enabled the legislators to think, speak, and act authoritatively. The first task for environmental lobbyists, therefore, was to be informed and reliable, and this task occupied their main energies. Few environmental organizations could encompass expertise on very many issues. Hence one of the major directions in the evolution of environmental lobbying was the creation of specialized groups, each of which became expert in one or a few subjects.

Lobbying also required close relationships with citizens. Much lobbying in the past had been from organizational leaders to legislators, but environmental affairs involved a citizen-contact dimension that required each group to develop a citizen constituency on which it could rely. Hence much environmental lobbying was devoted to mobilizing support from the wider general public, from professionals, and from sectors of the business community who happened to find the particular environmental objective advantageous to them. Environmental organizations that concentrated on lobbying came to the conclusion that success depended upon changing an environmentally unfriendly legislator to one that was environmentally friendly. As early as 1970 the League of Conservation Voters developed annual summaries of votes by members of Congress to encourage citizens to support candidates who had a stronger environmental voting record. In time, several organizations that were free to do so began to endorse and support candidates.

Administrative lobbying was quite a different affair. How the agencies carried out the requirements of the law was not predictable. Two hurdles had to be overcome: (1) the regulations to implement legislation, and (2) the actions to implement the regulations. At each of these stages the agency could carry out legislative mandates with greater or lesser effectiveness. Environmental organizations participated mostly in agency rule making because this was a relatively open process within which they could exercise some influence, but also because it involved a single point of decision making. Implementation required far more extensive "hands-on resources," which was beyond the ability of national environmental organizations to finance.

Legal action through the courts began in the 1960s and continued steadily thereafter. Environmental legal organizations evolved to take up litigation to force agencies to implement the law. Many cases were highly publicized and gave the impression of a vast amount of environmental litigation, but statistics of court proceedings made clear that it was a relatively minor amount. At the same time, the bulk of environmental litigation resources and action came from the regulated business community rather than the environmental organizations. Over the years, the courts constricted the ability of environmental organizations to use legal action by tending to confine it to procedural rather than substantive matters. When environmental issues involved technical questions, and many of them did, the courts tended to avoid making technical judgments themselves and to defer to agency expertise. Procedural issues often were more clear-cut. The legislation required the agency to take a given action by a given time, and when it did not do so, it was subject to court decision to require it to act.

While lobbying required knowledge of issues and lobbying skills, administrative politics required technical knowledge and expertise. Environmental administration entailed complex technical data and its interpretation. National environmental organizations often were the only environmental groups that could afford the cost of professional expertise, just as they were often the only ones that could afford the cost of legal action. Hence environmental groups were quite limited in their ability to enter into administrative politics as fully as they did legislative lobbying. Over the years these political abilities increased, but the progress was slow, and environmental organizations were continually outpaced in these resources by their opponents in the business community and in public agencies.

In the wider field of environmental implementation, the environmental organizations had limited capability. Most implementation, of course, was in the field, involving cases of specific environmental conditions such as air quality, water quality, land use, particular species habitats, toxic chemicals in a given place. Effective implementation, therefore, depended first of all on whether sufficient resources could be mobilized by people in a given locality to carry out environmental objectives in that place. Even where organized activity to deal with such local circumstances did occur, it often required technical skills that could be supplied only by some larger state or national environmental entity. At times national legal organizations could assist in enforcing permits, and this was done especially with respect to water quality. Land-use issues such as zoning, on the other hand, were usually subject primarily to local jurisdiction; at the other extreme, local land-use issues were sometimes taken out of the hands of municipalities and subjected to exclusive authority from state or national governments.

In all of these activities, information and knowledge were crucial ingredients and the widespread search for understanding linked them. Environmental organizations, national and state, were deeply involved in collecting and distributing information to enhance the level of environmental understanding. Citizens were eager to acquire relevant issue information, and organizations were well aware that such information was essential to foster the active public that would enable environmental policies to succeed. The first regular newsletter about environmental issues in the nation's capital was issued by the National Wildlife Federation in 1963 as it sought to establish better contacts with its expanding membership. The Sierra Club issued its first Washington newsletter in 1970. Most national organizations also issued magazines of various kinds with feature articles on a wide range of environmental subjects. Together they constituted one of the major outlets for investigative reporting on environmental issues. Eventually, electronic information provided users with timely information services about the course of public issues.

The subjects taken up in these publications were varied, complex, and extensive. One area was scientific knowledge; few environmental organizations were capable of conducting scientific research, but they fostered with some success the ability to keep in touch with and ferret out scientific work conducted by others to bring it to the arena of public decision making. Some organizations, such as those in the field of pesticides, were devoted exclusively to such a task. Often environmental organizations were involved with locating information in the hands of administrative agencies that those agencies did not wish to reveal to the public. Another area of information was about the way the political process worked at all levels: legislative, judicial, administrative, and executive; national, state, and local. One could argue that the environmental organizations constituted a major civic education enterprise to enhance citizen knowledge about the political process and to learn where and how they could exercise influence within it. One of their most significant, though little emphasized, activities was the publication of citizen handbooks, each one bringing together knowledge about the legislative, administrative, scientific, and technical aspects of a given issue.

Dissemination of information about environmental activities was slower to evolve at the state level and came to be one of the major objectives in the development of state citizen organizational activity. Most state legislatures and administrative agencies acted with far less openness than did national ones. Hence, information about state environmental affairs was often much more difficult to acquire and disseminate than comparable information at the national level. At the same time, few state environmental organizations had the ability to extend their activities widely in the field of information. A few developed state issue handbooks, a number produced legislative voting charts to guide citizens who wished to vote according to legislative environmental records, but most state organizations that focused on lobbying did not have the resources to conduct focused in-depth studies of state-level issues.

Such information activities occurred at levels below the federal government primarily through regional environmental coalitions that brought together state, local, and national organizations. Collectively they could, at times, develop sufficient resources to serve as more adequate sources of information. The Greater Yellowstone Coalition, for example, sponsored an annual symposium on science in the Greater Yellowstone ecosystem and served as a forum in which scientists as well as policymakers and citizens could consider and debate the relevant science. Another with more formal state and federal governmental involvement was the Chesapeake Bay Program, whose science and policy, citizen and governmental support and action provided a major example of systematic information development and dissemination. Federal programs that focused on understanding and improving water quality in estuaries facilitated this process. Similar information activities were closely connected with land conservancies that often brought together citizens and experts around a common interest in increasing the fund of knowledge about the areas in which they were particularly concerned.

THE PROBLEMATICS OF ENGAGEMENT

Citizens reached out in many directions to fulfill this desire for information. A range of institutions and agencies were available: federal, state, and local governments; the media; the business community; and environmental organizations. In this effort, polls showed that the public considered the environmental organizations to be the most reliable sources of information, much more so than either government agencies, the media, or the business community. A varied set of opportunities for environmental engagement had evolved, and those fostered by the organized environmental community were the most influential with the wider public.

Yet it was also clear that there were significant limitations to this facet of environmental engagement. Knowledge acquired and disseminated from the environmental organizations was focused overwhelming on the issues with which they were concerned. This gave a sharply practical focus to information, its acquisition, and its dissemination. Missing was attention to the larger context of environmental affairs. Forums for the discussion of larger environmental meaning, such as patterns of change over the years or the complex context in which issues were imbedded, did not evolve.

Amid such a vacuum the heavily ideological and symbolic perspective fostered by the media seemed to dominate. Hence the significance of environmental science, environmental economics, environmental values, or environmental knowledge did not come to front and center; neither did the role of environmental affairs in the evolution of the nation's constitutional system in its intricate relationships among legislative, judicial, executive, and administrative branches, or among federal, state, and local governments; and neither did the ability to broaden an understanding of events, usually to the social, economic, and governing context in which environmental affairs were embedded. The more limited media perspective, therefore, dictated the direction of public thought and had considerable influence with the nation's institutional leadership. Because the environmental community had no ability to present alternative perspectives in similar arenas, this one-sided state of public knowledge placed environmental affairs at considerable disadvantage in the public arena.

From the side of public understanding there were equally significant limits in the relationships between environmental understanding and environmental action. One of these came from the role of aesthetic engagement with the environment. Environmental aesthetics took many forms: personal enjoyment derived from visual appreciation reflected in wildlife magazines, books, and television programs; the tapes of environmental sounds; the development of environmental art; and environmental circumstances as the subject of mystery stories. Some involved public aesthetics, such as the nature symbols used on state license plates or the attempt to identify states with particular wildlife forms, as with the Kirtland's warbler in Michigan. Yet only on occasion did aesthetic engagement lead to political engagement; it was quite possible for individuals to participate fully in the one and to avoid the other.

From the standpoint of general environmental ideas, a more significant problem was the slow grasp of understanding about environmental limits on the one hand and human pressures on those limits on the other. A sense of environmental limits pervaded environmental culture at a fundamental but only vaguely felt level. Such a view was reflected in the picture of the Earth taken from space, which became a popular symbol in representing environmental affairs. At times this was given concrete and significant illustration, such as when the ozone hole or global climate changes seemed to bring to the fore that there were limits in the ability of the atmosphere to absorb gasses, or that there was a finite amount of air in which to absorb pollution from fossil fuels, or that soils and waters had a limited capacity to absorb and buffer the acid rain from the air, or that some problems such as lead in gasoline required that a physical limit be placed on the resulting emissions.

Although there was acceptance of the notion of limits in a few practical environmental circumstances, there was little evidence of a tendency to apply environmental limits generally, and efforts to bring that problem to the fore often fell by the wayside. Most environmental organizations themselves found that efforts by their members to focus on the "population problem" in the United States, a style of warnings posed by biologist Paul Ehrlich, became so controversial that they were dropped. At the same time, the problem of national consumption levels was talked about, but it rarely received attention as one of the nation's major environmental problems. Environmental engagement, therefore, involved only limited and often conflicting pieces of this complex of issues relating to growth of population, production, consumption, and waste on a finite amount of air, water, and land and did not extend to its larger meaning.

9 THE ENVIRONMENTAL OPPOSITION

One of the most curious features of contemporary environmental analysis is the limited focus on the environmental opposition. A wide range of literature exists about the organized environmental movement, written by those who associate themselves with it, those who oppose it, and those who view themselves as relatively neutral. But the environmental opposition as a subject for writing is rarely encountered. In their news stories, the media usually assume that there is an environmental opposition and stories about it appear as a momentary matter, but there is little systematic observation of the opposition as a persistent development in American society and politics of equal importance to the environmental impulse itself.

Analysis of the environmental opposition must consider three large questions. First, what are the roots of that opposition in the society at large, its economic, social, and political sources? Second, what are the various forms that the opposition takes in terms of arguments, policy alternatives, ideas, and action strategies? Third, what are the stages through which the opposition evolved from the beginnings of the environment as a subject of public debate and action to the present, and how has it waxed and waned over the years? These questions follow the general focus of environmental relevance outlined from the start of this book: political context and change over time.

SOURCES OF OPPOSITION

There are two major sources of opposition to environmental objectives. One involves the defense of older economic, social, and political cultures that are deeply rooted in the American past and for whom new environmental values represent a threat to the old. The other is rooted more in the contemporary economic interests that consider environmental objectives to be a major restraint on their activities.

The first of these is based on the long history of the American economy in farming and grazing, lumbering and mining, elements that were dominant in the nation's past but have decreased in importance in the face of new stages of economic activity. Associated with these are the manufacturing industries that process and refine raw materials into finished goods such as iron, steel, wood, and agricultural commodities. Associations representing these economic sectors are the most vocal and active in organizing and sustaining the environmental opposition. They represent past economic activities that have had difficulty accepting newer values and newer science, technology, and production practices.

This branch of the environmental opposition is closely associated with past cultural and social values as well as immediate interests for whom environmental objectives involve restraints on their economic activities. The extractive economy is embedded in the psyche of many communities and evokes both personal values and rewards placed on physical labor in extracting resources from the Earth. It shapes the outlook of large sectors of society such as the mining, wood production, and agricultural fields, which, as they decline relative to the growth in other sectors, produce defenders of their "way of life." Much of the contemporary influence of these past cultural values stems from their nostalgic character for Americans, who look back upon a past that they romanticize but from which they are divorced in the real world of day-to-day social and economic life.

This segment of the environmental opposition is strongest in those regions where the extractive economy is still viable or plays an influential symbolic and ideological role. Such opposition is relatively weak in areas that have long since passed through the extractive economy and the economy of heavy manufacturing into a more service- and information-based economy. In the daily lives of people in these areas of the country, brain work succeeds physical labor, and amenities—among them environmental quality—become increasingly important as major elements in their living standard. Regions of stronger environmental cultures are those of New England and nearby New York and New Jersey, the Upper Great Lake states, Florida, and the West Coast. Regions of weaker environmental cultures that are heavily shaped by former extractive industries include the western Gulf states, the Plains states, and the Rocky Mountain states.

The Rocky Mountain states currently display the most dramatic internal regional contest between the old extractive culture and the new environmental culture. The older economies of that region, such as mining, lumbering, and grazing, have declined steadily over the past several decades as newer stages of economic development have grown. The recreational and leisure enterprises have now become central to their economic well-being. Yet self-images die hard, and as the statistics about the changing economies of Montana, Utah, Arizona, and New Mexico come to the fore, they sharpen the conflict between the old myths and the new realities. Opinion surveys make clear that this region has one of the most favorable attitudes toward the natural world of any region in the nation. Yet it maintains a politics in which the old "lords of yesteryear," as one writer has called them, still play a powerful role.

Defense of the extractive economy was the initial element in the environmental opposition of recent years known as the "wise use" movement. Though the movement expanded its influence somewhat to the east and south, it remained most firmly based in the West. There the corporations in the grazing, timber, and mining industries, which now have far fewer exclusive claims on the public affairs of those regions, have rallied workers and others associated with those industries to shape a highly organized environmental opposition. The movement has had short-term success, beginning with the "Sagebrush Rebellion" that President Ronald Reagan sought to implement, but it is rooted in a fading older western economy.

A very different segment of the environmental opposition stems from the expanding manufacturing industries of the twentieth century, especially the chemical industries, which consider chemical regulation a threat to their enterprise. Many sectors of the chemical industry have, over the years, become involved in vigorous action to challenge environmental objectives. Some objected to the first stage of water-pollution control, with its attempts to reduce biodegradable discharges to streams, and others disputed regulation of air pollution. In the 1970s, however, the emphasis began to shift to persistent toxic chemicals, which ushered in a new phase of chemical-industry opposition.

This form of the environmental opposition has differed from that associated with the extractive industries in that it was confined to the industries themselves and had limited popular support. The industry challenged exploratory science that was working to identify the impact of chemicals on humans and the environment and tried to undermine that science with its own counterscience. The challenges are organized through general associations such as the Chemical Manufacturers Association or more specialized groups such as the Synthetic Organic Chemical Manufacturers Association. But they have also come together in specialized activities such as the American Industrial Health Council that takes the lead in challenging chemical regulation and the Chemical Industry Institute of Toxicology that supports chemical research to challenge the research on which regulatory activity is based.

Land developers constitute another major segment of the environmental opposition. The broad public interest in maintaining a more natural environment amid development applies to many lands in many settings: urban, suburban, and rural areas, and even prairies, deserts, and forested wildlands. In contrast with the extractive industries, they are more associated with the growth of urban and metropolitan areas. In each setting, developers seek to turn more natural areas into more developed land.

Over the course of the 1980s land developers became increasingly hostile to environmental objectives. The earliest of these had to do with zoning; environmentalists sought to modify the environmental consequences of development and persuaded many local governments to take action to require developers to absorb some community costs of development such as roads, schools, social services, police, fire, and parks. Such moves to control development gave rise to demands by developers that the public pay for direct costs and diminished the profitable value of property resulting from such regulation.

Other specific examples of the environmental opposition included the objection of container manufacturers and workers to container deposit legislation; the opposition of utilities to attempts to reduce both sulfur and nitrogen oxide emissions; the opposition of the agricultural chemical industries to efforts to restrain the use of pesticides; the resistance of the manufacturers of coke to attempts to control emissions from coke ovens; the opposition of wood-product manufacturers to new goals in forest management that might restrict timber cutting in favor of recreational uses and protection of wildlife and plants other than commercial trees; and the opposition of pulp manufacturers to adopting technology that would eliminate dioxin discharges in their wastewater. Environmental objectives involved the associated desire to modify many practices in a wide range of economic activities and, in turn, gave rise to resistance from those involved in such activities. As environmental objectives moved forward to constitute a permanent presence in the landscape of public affairs, so also did the environmental opposition.

During the last decade and a half of the twentieth century, one increasingly heard the argument that fundamental divisions in American society and politics over environmental issues had declined to insignificance. "We are all environmentalists," or so the argument went, with the special emphasis that now that regulated industries had "seen the light," they no longer opposed environmental objectives. Such arguments are wide of the mark; evidence about the substance of environmental affairs demonstrates that issues and attitudes about them remained much the same in the year 2000 as they had been in the 1960s. Environmental objectives remain a well-supported desire on the part of the American public, and the environmental opposition remains a major and relatively successful attempt to restrain them.

FORMS OF THE ENVIRONMENTAL OPPOSITION

To understand the various forms of the environmental opposition is especially difficult because of limited evidence. To analyze environmental organizations is relatively easy because their activities are readily made public, hence are easily observed. But the activities of the opposition are far more concealed from view. The range of exposure varies. More exposed actions, such as legislative politics, are easier to assess than are the less exposed actions such as administrative decisions or technological performance in production.

The most readily available evidence about the environmental opposition comes from its role in national policy activity. Here, though the opposition prefers secrecy, such as behind the protective device of executive privilege in agencies such as the Office of Management and Budget, it often is forced to make its views and tactics known through testimony before legislative hearings. The media are often tempted to probe on their own into the legislative opposition, so that, for example, in the heady days of the Republican control of the 104th Congress, the more obvious forms and the intensity of the environmental opposition became public. And the opposition often exposes itself through its financial contributions to legislators and its massive media campaigns to shape public perception.

From evidence of this sort one can conclude that the direction and intensity of the opposition has not changed much over the decades of the environmental era since World War II. It has come to accept some innovations in policy, but in such matters as air- and water-quality standards or nature protection, these often seem to be only temporary and opportunistic responses, which later give way to more fundamental opposition when opportunities arise to reverse policy. Similar opposition in regulatory rule making involves attempts to weaken legislation through regulatory modifications. While the regulatory process is relatively open and observable, most rule making is quite detailed and requires time and effort to follow. Hence most observers, whether the media, academic analysts, or historians, usually find these opportunities too burdensome to take up and miss the drama of the environmental opposition in administrative politics.

When rule making is turned into implementation, the activities of the environmental opposition are almost impossible to follow. Most environmental implementation involves direct, face-to-face relationships between the regulators and the regulated. This can be explored through the permitting process if one wishes to take up the task of working through a multitude of permits. But on the whole, observers forgo this opportunity because of the time and effort it takes. Public discussion of the relationship between the regulators and the regulated veers persistently toward ideologies or the general wisdom of "government" versus "market forces," a focus that sorts out ideological positions but hardly reveals the continuing interface between those who represent the "public" and those who represent the "private" components of that relationship. Hence this opportunity to ferret out the environmental opposition usually falls by the wayside in favor of the more general temptation to make vague references to "command and control" and "market forces," which convey little behavioral meaning.

In the mid-1990s a new group arose that provided more evidence about these relationships. It was called Public Employees for Environmental Responsibility (PEER), and it provided more opportunities for employees of public environmental agencies to provide evidence about the adequacy of environmental enforcement. In some cases, PEER defended whistle-blowers and publicized actions taken against them. In other cases, they surveyed the attitudes of environmental employees as to firm or lax enforcement and reported these in the media. In still other cases, they gathered comprehensive information about environmental permitting and followed the way in which applicants were treated favorably to the detriment of effective environmental programs. The work of PEER brought into fuller visibility the give-and-take between regulators and the regulated and enabled one to draw more factually supported conclusions about that relationship.

The role of the regulated industries in environmental science is also fairly easy to follow because the various choices made by both the regulators and the regulated are presented in the documentary evidence. In this case the task is one of being able to go behind the general ideologies of science, such as "good science" and "bad science," to describe the varied choices made by the disputing parties. For many years, for example, one issue in the health effect of potential carcinogens has revolved around how one counts tumors in the tissues of animals subjected to toxicological experiments, and this, in turn, depends on how one distinguishes tumors that are malignant from those that are not and whether one lumps them together when counting the evidence. If one wishes to pursue such matters of scientific dispute, some evidence is usually available, but most observers outside the relevant scientific fields shun the task because of their lack of knowledge of the points at issue or because of its detail and complexity.

In two major fields of environmental opposition the evidence is far less available because it is more fully controlled by the opposition: public relations and performance. Public relations evidence is readily available because wide dissemination is its major purpose. But though this evidence indicates what the opposition wants the public to know, it provides little opportunity to match public relations assertions with performance. All that can be done is to describe public statements as an integral form of environmental opposition, observe the way in which the opposition seeks to formulate issues, and then determine if that formulation is accepted in public debate.

Environmental performance provides even fewer opportunities for analysis, since verifiable information is almost totally absent. What is the actual level of emissions during the course of industrial processing? Regulatory agencies in two states (Massachusetts and New Jersey) have required more precise measures of performance by establishing points along the line of production, from beginning to end, that can be measured and reported. But these are in their infancy, and although they hold promise for some level of independent observation and verification of performance, especially for the crucial point of source reduction, that promise has yet to be realized. The regulated industries are quite aware that their own statements of performance are suspect. But in the face of their desire to increase public confidence, such as in the widely touted "responsible care" movement within the chemical industry, there is continual internal controversy between those who fear independently verifiable information and those who argue that independent verification is essential to establish greater public confidence. Until some system is developed to provide independent verification of environmental performance, this crucial realm of the environmental opposition will go uncharted.

To the eager observer, the most attractive form of the environmental opposition has come with the "wise-use" movement. Because that movement has some elements of a popular base and because the extractive industries play such a large role in it, the wise-use movement provides much grist for the media mill. At the same time, the highly ideological flavor of the wise-use movement and its strategies for manipulating the media through ideology render it relatively easy to monitor through its own documents and statements, and hence it becomes an easier focal point from which to chart the opposition.

EVOLUTION OF THE ENVIRONMENTAL OPPOSITION

Environmental opposition has grown in tandem with, and in many respects far more extensively and effectively than, the environmental movement itself. As new environmental issues have taken the spotlight, new sources of opposition have come into being. In its initial stages, the opposition was limited to piecemeal action, but as separate streams of protest grew, they coalesced into more comprehensive forms of anti-environmental action. At various times in the history of anti-environmentalism, organizations formed to advance more general economic objectives, such as the U.S. Chamber of Commerce, have taken on the task of opposing environmental programs, but new organizations have also arisen to oppose specific environmental objectives. Over the years these separate efforts merged to generate a powerful political movement to resist environmental advances.

One of the first organized forms of environmental opposition came from the nation's paper manufacturers, who in the 1950s formed the Forest and Stream Association to combat efforts of many a community to protest against air and water pollution as well as noise and smells from the industry. The paper industry formed its own association to take up joint political action to combat regulatory efforts. At first the action occurred at the local and state levels, but as the move for water-pollution control moved into the federal arena, opposition from the wood-products industry became more national in scope. As one of its major political strategies, the industry sought to carry out studies to show that pollution control was scientifically unsound, technologically unfeasible, and economically disastrous.

In the 1960s the growing interest in air pollution brought a variety of industries into the environmental opposition, especially the iron, steel, electric utility, and automobile industries. These industries attempted to influence policy at the state level by dominating the regulatory commissions and agencies. As national policy emerged in the 1960s, these industries focused especially on the task undertaken by the U.S. Public Health Service to study the effects of air pollution on community health. Known as Criteria Documents, these studies at first were carried out by the Public Health Service staff alone, but when the first such draft document, dealing with sulfur dioxide, was prepared in 1967, it became a target of attack by the polluting industries. In the 1967 Clean Air Act, Congress required that the document be revised under the guidance of an advisory committee that included representatives of the affected industries. From this point on, the scientific assessments of the health and environmental effects of pollution came to be a central target of the environmental opposition. Among other tactics, it devoted an increasing amount of resources to generating science to counter the exploratory science that underlay environmental objectives.

The emergence of toxic chemicals as a central environmental issue in the 1970s energized the chemical industry to become a leader in the anti-environmental movement. Passage of the Toxic Substances Control Act of 1976 and the increased focus on hazardous waste sites were major events in this development. But there were many background factors, such as the earlier publication of Rachel Carson's *Silent Spring* and the gradual emergence of workplace toxic-chemical exposure issues, which led to the implementation of the Occupational Health and Safety Act of 1970. In fact, it was a series of actions by the Environmental Protection Agency and the Occupational Safety and Health Administration (OSHA) with respect to cancer that provided the impetus for new major anti-environmental organizational efforts by the chemical industry. EPA and OSHA sought jointly to develop "generic cancer" policies that could be applied to the regulation of chemicals in general, moving the problem of regulation away from a "chemical-by-chemical" approach. Both agencies set forth, in quite similar fashion, the tests that could be applied in the case of each chemical to be considered. The EPA policy dealt primarily with pesticides and hence aroused only the Agricultural Chemicals Association, but the OSHA action affected many industries that now viewed OSHA as a major industry-wide threat.

The industry response took several forms. One was the formation of the American Industrial Health Council, which became a major political arm of the chemical industry in fighting the regulation of chemicals used in industry on every front. A second was the organization of the Chemical Industry Institute of Toxicology, located in Research Triangle Park in North Carolina, to undertake research on toxic chemicals. This provided the science that industry could use to combat scientific knowledge about the environmental workings of toxic chemicals, especially in terms of human health effects. This anti-environmental strategy seemed to arise automatically as every new frontier in the adverse effects of toxic chemicals was penetrated by environmental science. Still a third form was the American Council on Science and Health, a private organization led by Elizabeth Whalen that sought to mobilize public opinion against attempts to control toxic chemicals by seeking to undermine the credibility of the science underlying regulation.

During the 1980s a long-standing set of environmental opponents that had arisen in response to public land issues became a part of the increasingly cohesive environmental opposition. These included the forest, mining, grazing, and irrigation industries, each of which was organized in its own trade association and tended to fight its own battles over public land issues. Controversies over public land policies tended to focus on the west, but they also involved conflicts over eastern public lands, both federal and state, that were fought out more quietly. As public land and water policies placed more emphasis on the environmental and ecological importance of those resources and began to demand innovations in behalf of those values, the anti-environmental interest in commodity development also grew to oppose each environmental advance in succession—wilderness, wild and scenic rivers, outdoor recreation in general, species protection, in-stream water uses, and water quality.

A host of land-development issues brought land developers into the national anti-environmental movement. Formerly, they had been faced only with local attempts to modify their development plans, and often in terms of helping to pay for the associated costs of development such as streets and highways, which otherwise would fall on the taxpayer. But as environmental objectives focused more on such environmental consequences of development as the loss of open space, the destruction of wetlands and the resulting downstream flooding, or the adverse effects on plant and animal species, the development industries began to demand that the costs incurred by regulations on behalf of those objectives be paid for by the public.

The environmental opposition found more support within the Republican than in the Democratic Party. In the U.S. Congress in the 1970s and 1980s, voting support for environmental-improvement issues among the Democrats was about two-thirds in favor with a one-third opposition, but among the Republicans the balance was the opposite, about one-third in favor and two-thirds in opposition. The party balance at the state level was somewhat similar, with Democrats being the major vehicle of environmental improvement and the Republicans, of environmental opposition. After 1994, however, Republican Party toleration of its environmental minority shifted to a full-fledged attempt to submerge environmental sentiment within its ranks.

This move came about in close tandem with accelerated anti-environmental drives within the business community. Business opposition periodically was energized when it appeared that environmental objectives were on the point of obtaining more significant influence within the federal government. The first of these situations occurred during the Carter administration after the new president, as a result of commitments made during the campaign of 1976, appointed a number of environmental leaders to prominent positions in his administration. In a sharp response, the environmental opposition solidified its diverse elements to bring pressures to bear on the administration, and by the second half of the Carter administration, had succeeded in significantly muting that administration's environmental policy. The second occurred with the Reagan administration, during which the environmental opposition found widespread acceptance within its ranks and sought to reverse many environmental gains of the preceding years. But environmental strength in the Democratic Congress was sufficiently strong to ward off many anti-environmental initiatives, even though it was not strong enough to continue significant environmental improvement. The successful drive of environmentalists to force the resignation of Secretary of the Interior James Watt, who represented the environmental opposition, was a small victory for the promoters of environmental objectives.

With the Democratic victory in the election of 1992, environmental leaders once again found opportunities through appointments to the Clinton administration. But those appointments energized the environmental opposition to another stage of effort and resulted in closer connections between themselves and the Republican Party. After the Republican congressional victories in 1994, this newly forged relationship between the anti-environmental movement in the nation as a whole and the anti-environmental pressures within the Republican Party became even stronger. The Party now became a major instrument of anti-environmental policy, and Republicans with positive environmental views were placed under considerable pressure to conform to a growing official anti-environmental stance by the party as a whole. The resultant drama, which played itself out in the 104th Congress, focused largely on efforts by a contingent of environmental Republicans to maintain a legitimate position for themselves within the party.

In the 104th Congress, the Republican Party fashioned a full-scale attack on environmental policies. It appointed anti-environmental industry leaders to legislative positions; it drew upon those sectors to write bills; it favored them in public legislative hearings; it denigrated the findings of environmental science; it defended and protected radical anti-environmental activities in the nation at large; and its leaders gave expression to a comprehensive anti-environmental ideology.

THE PERMANENCE OF THE ANTI-ENVIRONMENTAL MOVEMENT

By the 1990s it had become clear that the anti-environmental movement was a permanent feature of the landscape of public affairs. It had started in many and diverse ways, as separate segments of the economy had found their objectives challenged by the increasing emphasis on environmental values. In the 1950s and for some time thereafter, the environmental opposition had apparently not yet been convinced that the environmental impulse was more than temporary, and they still thought it could readily be turned back by appropriate public relations strategies. But such was not to be the case; though environmental affairs still had a long way to go to become as integral a part of public affairs as were economic development, sports, and popular culture, they clearly constituted an influence that the affected industries would have to deal with permanently. Hence the opposition continued their efforts to perfect their anti-environmental political strategies.

The permanence of the environmental opposition contradicted the continuing argument from the affected industries and from the regulatory agencies that although those industries had at one time been environmental opponents, they had now "seen the light" and were "green." This argument was far more a public relations venture and part of the opposition's political strategy than an argument of substance. For though the opposition sought to make such arguments that they were now on the right track and that the environmental future would be assured in their hands, they continued to exercise vigorous opposition to environmental objectives in almost every quarter. Some of the larger corporations sought to help land conservation organizations, such as the Nature Conservancy, to acquire land for permanent protection, but their support was not matched by advocacy of public land conservation policies.

Amid these realities of the political world it seems rather fanciful to believe that the controversies were all just a matter of momentary tension or lack of understanding and that negotiation in the form of "environmental mediation" would solve the disputes. On this score, the past can well inform the future. Unless either or both the environmental movement and the anti-environmental movement suddenly go away, we face permanent controversies between competing values and institutions that will continue for the extended future.

As knowledge about and experience with environmental pollution proceeded over the years, and as controversies over its extent and proposals to curb it also proceeded, the context of "nature" and "natural" took on new meanings as it became an integral part of the environmental opposition. That opposition argued that environmental advocates overidealized nature and did not recognize both its dangers and its values. In these arguments the role of nature took on many new twists.

The environmental opposition argued that human pollution did not present a problem at all because such pollution was, in fact, inherent in nature. At times this took on the form of a media-style sound bite, such as when President Ronald Reagan popularized the argument that air pollution came from trees. But it also took on a more serious and widespread form as scientists and industries took up the argument in many more concrete cases.

This argument that what was thought to be human-induced pollution was only a part of nature seemed to arise with every specific pollutant. Thus, radiation was only natural and human-caused sources such as atomic explosions or nuclear power plants added only insignificant amounts to human exposure. Dioxin, so it was long argued by the chemical industries, was a natural product of burning and especially of forest fires. Industrial chemicals thought to be cancer causing in foods were insignificant because many plants used as food contained even more cancer-causing chemicals in their natural state. And as the argument arose that synthetic chemicals known as "endocrine disrupters" were responsible for changes in the human hormone balance, it was claimed that such endocrine changes were only natural; chemical companies sought to emphasize the point by urging that the term "endocrine disruption" be dropped in favor of the term "endocrine modulation."

The problem of the relationship between the natural and the human also came to be a significant feature of debate over management of forests, grazing lands, and wildlife. In these cases, environmentalists often objected to the increasing intensity of forest management, expressed, for example, in the concept of "industrial forestry." In contrast, they advocated a more "natural" forestry. At times, this idea had some influence within the forestry profession itself, as for example when the practice of tree planting became increasingly expensive and "natural regeneration" frequently replaced it simply as a less costly practice. In response to environmental objectives that called for less intensive management and especially modifications in harvest practices, the forestry profession argued that clearcutting mimicked the large disturbances of fire and windstorms that occurred naturally. Similar arguments raged over the use of fire, which professional foresters had long sought to suppress but which environmentalists, following the general idea of the beneficial effects of "fire ecology," argued should be allowed to burn if natural and perhaps even to be set as "managed burns."

Some writers who sought arguments to counter environmental emphases on the harmful nature of human-made chemicals latched onto the idea that "nature" was both benign and resistant to human-induced change. The first statement of this kind was the so-called Gaia Hypothesis popularized by a prominent English scientist, James Lovelock. His contention was that the Earth was quite resilient and that the assumed damage to the Earth wrought by humans would easily be corrected. In any event, life of some sort, not necessarily humanity and its culture, would go on. Some environmental theorists who believed that "mother earth" was threatened found his argument to be attractive, but to most it seemed to be far removed from the more practical context in which they were interested.

A decade or so later, a journalist-writer in the United States, Gregg Easterbrook, took up the same argument in his book *A Moment on Earth,* which involved a variety of anti-environmental claims that pollution problems had all been solved and there was no need to continue the environmental drive. Human damage to the Earth, if at all, was only momentary, and the Earth would readily recover on its own without additional human action. Easterbrook's optimism concerning contemporary environmental affairs was taken up by the environmental opposition as a weapon against the environmental movement, but reviewers did not take up his larger argument about natural resilience, and his book had influence only as a weapon in the hands of those who were skeptical about the entire environmental enterprise.

10 THE POLITICS OF ENVIRONMENTAL IMPLEMENTATION

Once adopted, environmental policies turned into issues of implementation and these, in turn, evolved in almost every case into environmental management. Unless policies could be carried out through some governmental agency, rather than be left simply to the world of private action and controversy in which the issues would be decided through the courts, they required some ongoing institution to carry them out. Environmental management thus dominated much of the larger world of environmental affairs.

Management was not only an instrument of implementation; it was also a realm of political choice. Environmental laws inevitably left many choices to the administering agency, and those choices came to be as central and as strongly contested as the legislation itself. Should the legislative mandate be carried out strictly or loosely? Should the agency think in terms those affected might readily accept? How should the agency determine the facts of relevant science or economics or the possibilities of fostering a more socially acceptable production technology? As the range of environmental laws increased over the years, so also did the jurisdiction of management agencies and hence a larger range over which choices were made. In the first half of the century, the world of management agencies was dominated by land and river management: federal forest, range, park, fish and wildlife, and river development. In the second half, these land and water agencies became more elaborate and were joined by those that dealt with pollution, known as environmental protection agencies.

The ideology of most management agencies was that they were politically neutral, that they only implemented the law and it was the legislature where "politics" reigned, while in administration there was more detachment. The agency applied knowledge and reason, while the legislature dealt with broad public, often emotional, currents. Yet the real world of management involved an ever wider and deeper domain of political choice, often based on lack of knowledge, uncertainty, and personal judgment. It was here that intense controversy arose over the technological and economic aspects of choices, which had been debated in the legislature only in the most general terms. Now ensued the fight over what these meant in stark detail. From all this one could argue that legislative politics was only the warm-up to the more intense, the more long-lasting, and the more sharply divided issues the agencies dealt with. It could be said that environmental politics began after a law was passed.

The choices made by management agencies were not minor. If, for example, the U.S. Forest Service decided that a significant portion of the national forests should be devoted to recreation or the protection of plant and animal species or old growth, then that meant less would be logged. If the Environmental Protection Agency decided that there were adverse health effects from lead in gasoline or that the level of power plant and auto emissions needed to be reduced, then that meant that environmental technology firms would get a boost and the affected industries would incur an added cost. Although the legislature had provided general authority for these decisions, it was the agency that turned that general authority into specific actions. Literally hundreds and thousands of such decisions became of enormous interest to those engaged in environmental affairs and a wide range of governmental and business activities that were sources of the environmental degradation.

Around each management agency there quickly developed a circle of people interested in its decisions. For the most part, these were much the same groups that had sought to influence the legislation. Now, after the law passed, they simply transferred their attention to the administering agency. They knew that many decisions were yet to be made. Each competing group sought to achieve its own objectives, and the pattern of controversy that had emerged in general terms in federal legislative debate was now repeated in management politics but with a greater degree of specificity and greater intensity.

Management decisions went through two broad sequences of action, one involving rule making and the other permitting, often applied through the states that carried out federal law. At the state level, a similar two-step process occurred, one involving rule making under state law or under federal requirements and the other involving specific permitting of those requirements. Rule making involved more general considerations such as science, technology, law, and economics. Permitting, on the other hand, involved a host of more precise circumstances pertaining to specific areas of land, air, or water, or specific sources of environmental harm or benefit. Should this land be used in this way or that? Should this source of pollution be required to clean up this much or that much? While rule making usually involved a wide range of interests, specific applications had a more limited clientele, each with an interest in more limited circumstances.

The world of management politics had its own peculiar characteristics, which were different from those of the world of legislative politics. One was that management politics was more exclusively involved with details of science, law, technology, and economics. Hence while such matters had been brought to bear on legislation in a general way, now they required more concrete detail. Just because of this context, administrative politics was more removed from the general public and more confined to give-and-take among the professional experts. Participants in administrative politics found that they could not make significant contributions to decisions unless these were cast in technical terms and were justifiable in light of the most up-to-date knowledge. The political game played in administrative politics involved who had the best and latest technical information and how one could control the process of gathering and applying technical data.

Administrative politics involved a variety of strategies. One was to overwhelm the agency with technical information that would slow down or even halt agency action just because of the time and resources required to cope with that information. Because environmental issues were highly complex, it was not too difficult to do this if one had sufficient resources. Another was to use other governing institutions to override an agency's decision. One could challenge the decision in the courts; one could seek help in the executive branch to override the agency, an option that was more difficult to counter because it could readily be carried out within the secrecy of executive privilege; or one could go back to Congress for a legislative fix, that is, a provision in some legislation that would explicitly reverse the agency decision. A variant of that type of action was to obtain from Congress a decision simply to reduce or eliminate funding for that particular program. In ways such as these, agency decisions were not isolated from the larger world of environmental politics but played a major role in the complex of forces shaping the nation's entire governing system.

MANAGEMENT AGENCIES

In the first half of the twentieth century, the most significant environmental management agencies pertained to land and its resources. There were four main federal land agencies: the U.S. Forest Service, the National Park Service, the Bureau of Land Management, and the U.S. Fish and Wildlife Service. And there were many state agencies, such as state land commissions in the West managing "trust lands" for income-producing purposes to benefit education, state fish-and-game agencies, state park agencies, or state water boards. Over the years each agency had developed programs to manage a particular resource, and this had generated a distinctive set of state political circumstances. It was difficult to understand not only the complex of state agencies and the politics in which they were involved but also the intricate relationships between state and federal management agencies.

All of these agencies went through a new and profound set of circumstances after World War II. For each of them the natural resources they managed had long been thought of in terms of material resources, such as wood, forage, water, fish, and game. Now, however, the public began to look upon these resources as an environment for home, work, and play—uses they considered equal in importance to commodity production. Since a host of related technical professionals such as foresters, range experts, game experts, and hydrologists as well as a host of user organizations had been deeply committed to commodity development, they now felt threatened by the new environmental objectives. Federal laws expressed the new objectives: the Endangered Species Act of 1973, the Forest Management Act of 1976, the Public Lands Act of 1976, and the Fish and Wildlife Organic Act of 1997. Tense and unabated conflict between old and new continued to shape the environmental politics of resource-management agencies.

The National Park Service was the closest in tune with the new recreational and ecological objectives and hence adjusted most easily to change. The U.S. Fish and Wildlife Service, with its primary mission to protect and enhance the nation's wildlife, also had little difficulty in absorbing the new objectives and, in fact, came to be the most vigorous in advancing them; it was able especially to make the shift from an earlier emphasis on game and eradicating predators to appreciative wildlife uses. On the other hand, the U.S. Forest Service and the Bureau of Land Management, with their primary commitment to resource extraction and development, found the new objectives to be a major threat and took up a massive resistance to them that remained vigorous and effective to the end of the twentieth century. Water-development agencies, primarily the U.S. Army Corps of Engineers, the Bureau of Reclamation, and the Soil Conservation Service, all equally committed to resource development, had problems with the new environmental objectives and made adjustments only selectively and then only in limited ways.

New agencies arose, primary among them the Environmental Protection Agency, to implement new policies dealing with pollution. From their inception, these agencies were more closely associated with environmental issues and did not go through the painful transition from commodity to environmental objectives. At the same time, however, they confronted resistance to their efforts from both private and public agricultural and industrial sectors that were sources of pollution and that the environmental protection agencies were called upon to regulate. Hence the EPA, the central environmental agency of the federal government, found itself in political tension with a wide range of private and public institutions. It also found itself at odds with land-management agencies over such issues as forest management and stream erosion. Over the years the EPA gradually developed an ecological perspective that came to be an important, though often tenuous, feature of the wide concept of environmental harm that was formulated as "ecological risk."

Surrounding federal environmental agencies were several public and private institutions that shaped the agencies' larger political niche. Much of their activity involved the private sector, that is, private businesses and citizen environmental groups. The regulated industries presented the agencies with more continuous demands and commanded more of their time and attention; these relationships involved large amounts of scientific, technical, and economic expertise. Relationships with the environmental community were more sporadic, more involved with policy concerns and far less with technical communication. With fewer resources, citizen environmental organizations tended to focus on selected issues and therefore to engage the agency less frequently. Continuous agency-industry relationships often became highly influential in a subtle but powerful manner, shaping both specific agency decisions and how the agencies thought about environmental affairs.

How could one ensure that a development-mandated agency could also be an effective agent of environmental improvement? At first, the dominant way of resolving this problem was to share responsibility, which of course the resource agencies did not relish. When the EPA was established, two such knotty problems were dealt with by transferring the responsibility for protection to the EPA: protection against radiation from nuclear plants was transferred from the Atomic Energy Commission to the EPA, and the authority to regulate pesticides was transferred to the same agency from the Department of Agriculture. The pesticide transfer brought forth a massive hue and cry from the chemical and agricultural interests and led to a provision in the 1972 Pesticide Act that a special panel known as the Science Advisory Panel be established in the EPA to review the scientific aspects of all pesticide decisions. Its personnel was to be appointed by the EPA with the advice and consent of the Department of Agriculture. This ensured crucial influence within the EPA from the affected chemical and agricultural pesticide interests.

A variety of pollution-related issues in ensuing years involved the EPA in intricate interagency politics; two examples are pollution caused by U.S. military installations and the effects of forest practices on water pollution or biodiversity. Such involvement was fostered by the environmental impact statement required in the National Environmental Policy Act (NEPA) of 1969, which in its initial phase involved primarily interagency review. NEPA sought to resolve interagency disputes by requiring that each federal agency contemplating a major action should share its proposal with other agencies for their review. Although this enabled the agencies to iron out differences among themselves, it also involved the EPA especially in the environmental implications of other agencies.

Management agencies at both federal and state levels of government emerged in tandem, and their relationships were varied and intricate. In some instances, federal programs imposed requirements on the states; for example, federal pollution-control laws imposed minimum nationwide standards while giving states the freedom to establish stricter standards. The course of this balance between minimal and higher standards was influenced heavily by the business community, which persistently sought to prohibit stricter state standards than required by the national minimal standards. At other times, federal legislation prohibited states and municipalities from adopting their own standards stricter than the federal, as was the case, for example, with noise. Such federal preemption of state authority came from the demands of industries that feared local standards. The pesticide industry continually sought a similar style of federal preemption of state and local pesticide regulations.

Far more significant in state-federal relationships was the question of funds and scientific and technical expertise. Federal standards usually carried with them federal financial assistance to help carry out the programs. Assistance to construct municipal sewage facilities lasted in grant form for many years, but then was later shifted toward loans in which more state funds were required. The federal drinking water program went through a similar evolution, but in this case the federal requirement for monitoring local drinking water supplies came to be a major bone of contention because of its cost. Municipal water suppliers, both private and public, balked at absorbing the cost of monitoring, especially for toxic chemicals.

Federal-state management relationships also involved the federal response to state environmental initiatives. A number of states undertook their own environmental programs and then sought to generalize them through federal action. In air-pollution matters, Minnesota and New York took initiatives on acid rain, and Minnesota and Maine took initiatives on mercury; in pesticide matters, Massachusetts and New York prohibited the use of alar as a growth stimulant in apple production. California for many years took a leading role in measures to reduce pollution from automobiles, and in health matters, took initiative to restrict the use of toxic chemicals that had harmful effects on human reproduction. Throughout the states the protection of plant and animal wild resources seemed to move ahead more rapidly than at the federal level. A federal nongame wildlife law authorizing a federal grant-in-aid program to states was passed in 1980, but it was never funded and so came to naught.

States also depended on the federal government for scientific research and expertise. Some states developed their own environmental science capabilities. California was noteworthy in this regard, as were New York and Massachusetts. Federal resources were often used in combination with state resources to fund research, in forest and wildlife affairs for example, and to facilitate the development of research in state universities. As time went on, resources for advancing environmental knowledge came to be increasingly crucial. More complex issues required more extensive and costly investigation, so much so that among those who objected to environmental programs there was a tendency to argue that it was not more science but policy change that was needed. This was the case with wetlands and endangered species programs, which by the 1990s had developed their own extensive body of scientific knowledge that played an important role in policymaking.

THE POLITICAL WORLD OF MANAGEMENT

Environmental implementation and management focused less on broad policy questions that were more the province of legislation and more on the narrow technical details that turned general policy into on-the-ground implementation. Surrounding each management decision were issues about the scientific judgments, the technology chosen, the economic justification in terms of benefits and costs, and the decision's legal standing. To be a party to the relevant decisions one had to bring to bear knowledge and expertise in each of these areas. Agencies became major sources of expertise, and the groups that sought to shape agency decisions found that they had to have equally credible experts to be able to affect decision making.

Agencies also established effective relationships with those in the political world who sought to influence agency decisions. The problem was to do this in a manner that established "fairness," to make sure that one interested party did not have advantage over another. This was difficult, since it was invariably the case that other government agencies and the regulated industries had the funds to bring a high level of technical expertise into decision making, while citizen environmental organizations did not. Hence inclusion was inevitably selective.

Almost every field of agency decision making involved advisory committees, and these were usually of two kinds: one involving representatives of the various "interests" that stressed policy issues, and the other involving technical experts. Membership on these committees was a major issue of controversy. One of the most long-standing controversies involved committees used in managing western grazing lands under direction of the Bureau of Land Management, which were composed overwhelmingly of the cattle grazers. Advisory committees were the focal point of considerable political maneuvering to secure members representing one group or another, or one technical point of view or another.

Among the most crucial and controversial advisory committees were those that provided judgments about scientific issues. The first such judgments were authorized by Congress in the Clean Air Act of 1963 in the form of "criteria documents," which were compilations of the scientific work on particular pollutants. When the U.S. Public Health Service produced the first draft of the criteria document on sulfur dioxide in 1967, it was roundly criticized by the coal, oil, iron, steel, and auto industries on the grounds that its scientific judgments were defective. A provision of the 1967 Clean Air Act required that the document be revised under the guidance of an advisory committee that included representatives of the affected industries. This set the initial pattern for the politics of scientific assessment, which grew and became more elaborate over the years. Succeeding controversies over EPA assessments led to the creation of a separate agencywide Science Advisory Board to which the EPA was required to submit policy-related scientific issues for its judgment.

Scientific issues then came to involve disputes over the personnel of the Science Advisory Board as well as its own procedures and judgments. The Reagan administration sought to change the board's personnel so that it would reflect the views of industry-related scientists. This, however, led to an outcry of protest from the scientific world and was an important part of the discredit heaped on the EPA during the early years of the first Reagan term. When William Ruckelshaus was called in to head the EPA to help restore its credibility, one of his first tasks was to restore the credibility of the board. This was done by providing a balance in board membership more commensurate with the range of opinion within the scientific community and by appointing as chair a highly respected scientist who had the confidence of a broad segment of the scientific community.

The political role of science advisory committees was demonstrated rather sharply by the Science Advisory Panel in the EPA, which dealt with all pesticide-related science issues and was a body quite distinct from the Science Advisory Board. The panel was established by the Pesticide Act of 1972 as a concession to the pesticide industry when President Nixon transferred pesticide regulation from the Department of Agriculture to the EPA. The EPA appointed the panel but with advice from the Department of Agriculture. This meant that among its members were scientists friendly to the use of pesticides and skeptical about their harm to people or wildlife. The Science Advisory Panel continued to be a major source of political influence from the pesticide industry through the role of its scientists, who made scientific judgments friendly to its objectives.

Technology issues were equally compelling and revolved around the question of how to promote more environmentally desirable technologies. Such technologies were often well developed and available, but industry balked about applying them because it desired to realize continuing profits from older investments before making new ones. At times there were breakthroughs, such as in the case of vinyl chloride manufacture, in which new technologies led to the recovery of waste materials that more than offset the cost of recapturing them. Or when the 1990 Clean Air Act required that emissions of sulfur dioxide be reduced by 50 percent over the next ten years, this in turn generated markets that stimulated both cheaper coal and cheaper scrubbers and enabled utilities to control emissions at far lower costs than anticipated. The EPA emphasized the desirability of implementing technologies that were already in place in advanced sectors of industry rather than those that were only "on the shelf." The agency's efforts were aided by both the clean water and clean air laws that established technology models as standards to which industry would be required to conform.

Then there was the question of economics, the relative balance of costs and benefits, which involved several knotty issues. One was the relative benefit of a new technology in terms of jobs, income, and profits versus the relative loss to an old technology that would be replaced. Most of the public debate emphasized the losses rather than the gains, and many economists were prone to do likewise. Industry's estimates of costs averaged about double those of the EPA's. At the same time, there was continual controversy over the benefits, with the sources of pollution claiming them to be far less than the EPA maintained.

Such questions defined the world of administrative politics as one in which one set of technical professionals confronted another set to decide on matters in their particular specialty. Those who could bring expertise to bear on the issue had the upper hand in establishing both the terms of the issue and its resolution. The regulated industry had far more resources than did the environmental community, hence it was able to establish the climate of discussion and issue resolution. The relevant world was one of face-to-face relationships between the regulated industry and administrators. Relationships with the environmental side were more sporadic and thus less influential in shaping how management thought. This context became especially important in setting the terms of on-the-ground implementation through permits, monitoring to determine compliance, and supervisory actions. Agencies rarely had sufficient personnel to assess the results of regulations.

Amid these administrative circumstances, the agencies either sought the security of a quiet and relatively calm relationship with the regulated industry or risked controversy by pressing for more effective performance in order to advance environmental objectives. In this tussle, the environmental community endeavored to bring to the fore the facts of environmental circumstance, while the regulated industry attempted to reduce the requirements of environmental compliance. The agency could well think of itself as "in the middle" of competing forces, but the "middle" was shaped more by the regulated industry, with whom the agency evolved a common interest, rather than by the environmental community.

Decision Making

Debate over how decisions should be made played a significant role in administrative politics. Legislation was far more open than administration largely because legislators were elected and administrators were appointed. Hence legislators felt that they were somewhat more accountable to the wider public than did administrators. Administrators felt that they were "above politics" and that they responded more to standards and guidelines formulated in their own professional and administrative worlds. But agency actions clearly involved crucial decisions in which the public was much interested. As a result, demands arose for administrative decisions to be more open. The agencies, in turn, resented these demands on the grounds that they, the agencies, represented the public as disinterested decision makers.

The first demands for openness came from the regulated community and pertained to an issue of classic environmental significance: the effect of the use of methyl mercury in treating a fungus on apples. In the early 1930s, officials in the U.S. Department of Agriculture had decided that such a fungicide risked human harm and had banned it. The industry then argued that this action had been arbitrary and capricious and that it had had no opportunity to contest it. The debate continued for many years and was a major cause for the Administrative Procedures Act of 1946, which outlined how agencies were to function. Proposed agency actions were required to be published in the *Federal Register* and the public invited to respond.

As time passed, however, the setting for regulatory action involved many consumer-based issues, and it was from the public rather than the industry side of these issues that demands came for more openness and public scrutiny of agency decisions. Citizen environmental organizations played a major role in the evolution of these procedures. Often they were joined by the regulated community, which also had an interest in agency responsiveness. For example, when industry found the executive branch more friendly to their views, they became more interested in quiet influence within the agencies and the office of the president; when this was not the case, they fostered more openness.

One of the most widely known devices for open review was the environmental impact statement, or EIS. This grew out of the 1970 National Environmental Policy Act, which required federal agencies to analyze the potential environmental consequences of a proposed project before undertaking it and then to submit the plan to other agencies for review. In its original intent, these interagency reviews were to be kept confidential. But under prodding from both the media and environmental organizations, they were made public and provided citizens with some opportunity to obtain information as to the basis for agency actions and to challenge their adequacy. The EIS process was quite limited. It was expressly excluded from all water-quality permits and not used generally in pollution issues, but was confined more to development-related issues in water and land management. At the same time, the courts effectively eliminated the substantive requirements of NEPA and confined it to procedure. Once the agency had satisfied the courts that it had explored "all relevant" environmental factors, the courts left the agency free to make any substantive decision it chose.

The EIS process was one feature of a continual problem: how to make information available to those who might wish to have it? This was the main feature of what came to be known as "public participation." Agencies, often wary about doing this, varied from those more friendly to information access and those less so; they also varied according to the attitudes of the presidential and gubernatorial administrations. Many a citizen group, for example, found that publications coming from agencies during the Carter administration suddenly ended with the Reagan administration, or that individual requests for information were processed quite slowly. As technology advanced, new devices emerged that enabled agencies to process requests for information with relative ease. Publishing information on the Internet enabled each agency to make a major show of openness, but as one roamed the web sites one could readily make out the selective nature of this information availability.

In enhancing the flow of information the environmental community carried on a number of sticky disputes with the agencies. One involved the availability of documents that might be significant in evaluating agency decisions: would they be available and at whose expense? Would one have to go to the agency building some miles away to see them or could one obtain them by mail? Could they be photocopied and at whose expense? Another was the degree to which the agency could use information that was not subject to scrutiny before it was used to support a decision. The regulated industry often pressed upon the agency the "latest study" that showed that such and such a proposal was unwise. The courts ruled that if such information were used by the agency in its decision, the entire case had to be reopened, the new information included in the *Federal Register,* and opportunities for rebuttal permitted. Still another was the way in which the agency summarized the submissions it received as comments on its proposal. How would they be categorized and added up? How would one count up printed form postcards and standardized letters from the public in contrast with those individually composed? Agencies were prone to shape this process to justify their choice as to desirable action.

Procedural questions came to be supervised by the courts, and often they were more inclined to hold agencies to account for procedural deficiencies rather than substantive ones, in which they were more willing to defer to agencies on the grounds of their superior technical skills. In so doing the courts brought many issues into the realm of fair procedure under the elaboration of the Administrative Procedures Act of 1946. Two ideas played important roles in court decisions: one was "fairness" and the other was the closely related one as to whether decisions were "arbitrary and capricious." The courts shared a fundamental instinct that it was unfair for one party in a dispute to have greater influence with the agency than another; hence procedures, information, and opportunities to rebut information should be equally available to all. But there was an equally important view that the agency should weigh all information with fairness and not be "arbitrary and capricious," that is, it should be guided by the objective facts and considerations rather than impose its own less substantiated judgment. Since information deficiencies were rampant in environmental affairs, it was easy for the discontented to argue that insufficient information or the wrong information or the wrong scientific advisors were relied upon for the decision. Over the years the courts developed their own views as to what rendered a decision free from being "arbitrary and capricious" as well as what made it "fair," and how these went often played a significant role in determining the direction of environmental affairs.

Litigation played an important role in how environmental issues were decided, and here also procedure played a significant role. Early on, environmental organizations brought legal action to require agencies to be more open and responsive, and many a statute explicitly gave citizens the right to sue agencies. But these suits were quite limited in their scope; they were usually permitted only in the case of a "mandatory" rather than a "discretionary" agency decision, that is, only if the law said in plain language that an agency "must" do such and such rather than "may" do it, which implied that the agency was free to make its own choice. While the environmental community often sought the language of "shall," the regulated community preferred "may." At times the most stringent requirement of a statute was to require that the agency produce a regulation or a study by a certain time. When it failed to meet a deadline, it was subject to court order to do so. Agencies missed many such deadlines, and on occasion environmental organizations took up the resulting potential for court action.

State environmental agencies were subject to the same procedural issues as were federal. On the whole, they tended to be more closed and their strategies over the years were to restrict decision making so as to reduce citizen influence. At the federal level, for example, it was assumed that the interests of members of the public differed from the interests of agencies, hence opportunities for public representation on advisory committees should be provided. But in the states, one often heard that the agency represented the public and hence no special representation by a public interest was needed. At the same time, state governments exercised considerable choice in selecting representatives of the public, often choosing those that would be less likely to criticize agency decisions over those that would. State agencies provided far more representation by the regulated industries.

State agencies also provided far less information to the interested public. Only a few states had environmental impact requirements for agency action; only a few states provided information as readily as did federal agencies; and many placed barriers in the way of ready access to agency documents. More limited resources lay behind much of these more closed state agencies. Yet much of it also depended on the environmental culture of the particular region or state, and as a result, state administrative politics differed markedly from state to state. In fact, one of the more elusive aspects of understanding state environmental affairs is the more complex problem of grasping the close relationship between that climate and the resulting politics of environmental management.

THE ENDURING POLITICAL CONTEXT OF MANAGEMENT

Over the years the political circumstances of implementation shaped a pattern of behavior that defined environmental managers as cautiously conservative. They became painfully aware of the continual presence of those who vigorously wished to advance environmental objectives and the equal presence of those who sought to deter them. Lurking behind these immediate political realities, and often somewhat ominously, were the legislature, which at a moment's notice could call for a public hearing for some administrative action that an elected official did not like, and the courts, which might in some unpredictable fashion declare an action contrary to the law or the constitution. The political roadblocks to smooth implementation were many and frequent.

It was rather customary for the legislature to outline broad environmental objectives that became far more limited as action wended its way from passage of the law to shaping the general rules under which it would be administered to on-the-ground implementation. The agencies did not seem to be politically free to advance environmental objectives in the spirit of the law when they were challenged by the regulated community to adhere strictly to a narrow statutory meaning of its words. Citizen environmental organizations and the environmental scientists who charted new stages in environmental objectives and new knowledge pressed for the clear intent of the law. In the face of these contrasting pressures, management agencies were not neutral, but they also were aware of the risks they entailed from political retaliation. The close observer of the ensuing sets of dramas must continually search for the ways in which agencies sought to maintain some degree of freedom to act according to their own wisdom within a reasonable reading of the law.

Anti-environmental influences made the agencies cautious to make sure that their assessments of science, their technological proposals, and their economic analyses were supported by a sufficiently large body of professional experts and interest groups to provide them with the necessary political support to make a proposed regulatory action stick. The agencies were continually pressed by the regulated industries that could mobilize their own technical resources for their own views on scientific, economic, and technical matters. In this contest, the courts often played an important role. For while they often called the agencies into account for the potential economic burdens on the world of private enterprise, they also tended to defer to the agency on issues of the scientific adequacy of agency decisions.

Amid all these factors of science, technology, economics, and law, the agencies tended to carve out a politically workable ground fostered by their search for safer political positions. Although they sought to act along the general lines of the objectives of the law, they were prompted not to veer too far beyond the objectives of the regulated community. Administrative politics merely transferred the same complex of political pressures from the legislative to the administrative arena. Agencies especially tended to avoid innovations that might subject them to industry opposition, which relegated them at the most to cautious incrementalism rather than vigorous innovation.

In concert with the watchful caution exercised by the agency was the significant tenuous relationship between legislature and agency. The Congress realized continually that it could not make the detailed decisions required to implement regulatory statutes. Yet, just how far should it go? There were those who argued that the Congress did not have the technical expertise required and hence should give the agency considerable leeway to make its own choices to implement the general objectives of the law. But, on the other hand, there were those who argued that in doing so Congress gave too much discretion to the agency, which gave rise to the wrong decisions, thus it should give the agency more specific guidance. As those interested in the outcome of regulation focused more sharply on its specifics, there was a temptation to veer one way or another depending upon the outcome of administrative decisions.

There were specialists in government who argued confidently that this problem should be resolved one way or another. But they rarely faced the basic problem, the ability of decision makers—whether in Congress, the agencies, or the courts—to cope with the increasing complexity of environmental circumstances. They were all faced with the implications of the rapid growth of detailed technical knowledge. Amid those implications there were tendencies on all sides to defer to the "experts" in the agencies. Yet if the decisions of the "experts" did not turn out as one wanted, there was an equal tendency to undermine the legitimacy of the experts as political actors. It seems wondrous indeed that amid such political scrambling any persistent environmental progress at all could be made!

11 ENVIRONMENTAL SCIENCE

The steady development of environmental science is one of the more remarkable features of recent environmental affairs. As late as the 1960s environmental science was still in its infancy, but by the end of the century there was much vigorous activity in research, published articles, new journals, professional exchanges, and conferences. The central journal in the field of environmental pollution, *Environmental Science and Technology,* published by the American Chemical Society beginning in 1967, became by the 1990s the Society's journal of largest circulation save for its lead publication, *Chemical and Engineering News.* Publications in the fields related to the natural environment such as biology, ecology, geology, or hydrology appear in a variety of journals, some of pre-environmental vintage such as *BioScience,* but new journals such as *Conservation Biology* or *Natural Areas Journal* serve as more eclectic publishing outlets. A growing body of knowledge about the pathways and networks in the environment and how they work, or the effects of chemicals on both humans and the natural environment, has evolved into a large body of scientific data about the world in which we live. Taken as a whole, it is one of the most significant developments in science in the last half of the twentieth century.

Environmental science has a dual origin, one involving the evolution of science itself, from the impetus of scientists seeking to understand the world more fully and precisely, and the other involves the social context of science, the new public interest in environmental affairs. As citizens faced changing environmental circumstances, they turned almost instinctively to science to understand the biological world they sought to preserve and manage, to comprehend the sources, pathways, and effects of pollutants, and to get a grasp on the details of the impact of new development and population on their communities. The initial impulse in almost every environmental circumstance was to learn more about it in order to tackle it more effectively. One might well argue about the degree to which the drive to enhance environmental scientific knowledge came from the world of science or from the environmental public, but certainly the two reinforced each other to forge a persistent goal to understand better the world they sought to shape.

Yet this significant scientific development was, at the same time, deeply involved in controversy. While one might view the evolution of science in public affairs simply as a matter of applying knowledge, environmental science at almost every point became subject to intense disagreement. That controversy was due in part to the disputes among scientists themselves, which seemed to be an integral part of the way in which frontier knowledge that is complex and difficult to pin down with conventional scientific method tackles new areas. The more environmental science advances, the more specializations and their particular advocates grow in number, each of which has a specialized stake in the issues.

But in equally large part the controversy was due to the relevance of scientific knowledge for policy. The progress of agreed-on environmental knowledge was slowed considerably by the way emerging knowledge was challenged at almost every step by developers and industries for whom that knowledge might have regulatory implications. As a result, the relevant developers and industries sought to turn science in their direction and attracted scientists who could help with that objective. Environmental science became highly controversial because the affected developers and industries fostered that controversy in behalf of their economic objectives.

The result of these political involvements of science was that the field of environmental science seemed to develop under a cloud of suspicion. Many in public life thought of it not as one of those many dynamic elements of science that have arisen in the course of history when new frontiers are identified and then explored, but as something forced from extraneous and undesirable sources, diverting the search for knowledge into unproductive pathways that were not all to the good. As new environmental knowledge entered the fund of scientific understanding and then the field of public dispute, that frontier knowledge was often dismissed as irrelevant if not erroneous and labeled often with the stigma of "bad science."

Yet the evolution of environmental science continued, and a wide range of researchers found in their work a slow and steady process by which a world formerly unknown to them bit by bit became known, a scientific enterprise still only in its infancy and with a long course to run. A vast and complex world called the environment came to engage the imaginations of both scientists and citizens to explore and understand it, much in the same way as the exploration of space has caught the human imagination today. Amid small discoveries that pin down the known, a vast unknown environmental world remains to drive the human imagination toward further discovery.

IMPULSES IN ENVIRONMENTAL SCIENCE

Three impulses have shaped environmental science: exploratory science, defensive science, and managerial science. Each plays a separate and distinctive role in the scientific enterprise. Exploratory science carves out new knowledge to turn the vast unknown of the environment into the known. Defensive science looks with skepticism on the results and seeks to erect a counterweight to the new knowledge. Managerial science seeks to reconcile the two, develops syntheses of existing knowledge, and fosters new knowledge to bring some degree of workable agreement out of the welter of contention and debate. Because of the central role of political struggle in the entire process, that synthesis is heavily shaped by what is politically and legally acceptable.

Exploratory science proceeds in rather conventional and traditional ways. A scientist identifies a problem or a new question for research arising from some initial discovery; the results of the research lead others to elaborate the conclusions further to look at new evidence; a central finding is refined; new methods of investigation are applied. In this ongoing research, the results of one step define the next step, and the scientific enterprise is drawn on by human imagination, which generates further inquiries into what the subject is all about.

This new thing called the "environment" engaged the imagination of scientists not long after World War II. Soon environmental science became a driving force as each new piece of research revealed not only more understanding but more yet to be discovered. The enterprise generated a potential new field of knowledge that could absorb an enormous amount of human energy and resources. A huge environmental world, now barely known but with almost unlimited dimensions for discovery, became a major field of scientific exploration.

Soon, however, another force for advancing environmental science came into play: the objections of a number of scientists whose direction of thought was not exploratory but defensive. A lot of nonsense was being discovered, so these scientists maintained, and investments of time, energy, and funds were needed to set it right. This view was driven far more by the business community, which feared the potential regulatory implications of new science. Almost every field of actual or potential regulatory activity that involved the application of environmental knowledge generated a counteraction to question that knowledge or to develop further knowledge that might mute or even undermine its conclusions.

These scientists were driven not by the excitement of discovery amid new frontiers but by the overriding desire to set incorrect knowledge right. The timing of their research was guided less by stages in the evolution of knowledge than by the fear of its potential policy implications. Their stance in science was more negative than positive, more geared to disproving than to proving, more to arguing against than arguing for. Their way of thinking and talking about all this was to divide science and scientists into "good" and "bad," and they were as absorbed by trying to find out why "bad" scientists drew "bad" conclusions as they were by trying to establish an acceptable science itself.

A major effort in this defensive science on the part of the regulated industries was to make connections with scientists in the universities who were as skeptical of the new knowledge as they were. Often the significant steps in those connections came as the industries searched for professional experts to support them in court cases. Other steps involved allocating industry-generated funds to research, establishing connections with young scientists whose work was emerging, organizing their own conferences, and in general developing a counterculture of defensive science.

Still a third direction in environmental science was shaped by the administrative agencies, usually at the federal level, where resources existed to finance the work, such as at the U.S. Fish and Wildlife Service or the Environmental Protection Agency. These agencies were interested not so much in exploratory or defensive science but in establishing a successful scientific basis for potential administrative action. Their primary task was to assess the results of existing knowledge in order to apply it to policy. They were pressed to act by frontier-breaking scientists who felt that new knowledge should be applied more quickly, and they were equally pressed not to act by the regulated industries. The agency task was to mediate between exploratory science and defensive science by developing a workable agreement among a sufficiently large number of scientists to sustain its policy application. The heavy role of law and lawyers and the ever-present potential for legal action brought by the regulated communities on the adequacy of the science used for regulation led to an influential legal context for defining what was scientifically acceptable.

DIRECTIONS IN ACCUMULATING KNOWLEDGE

Environmental scientific knowledge accumulated in four different fields: (1) the ecosystem as the context in which the natural world works, and the relationship between change in the natural world and change resulting from human action; (2) the effect of environmental contaminants on human health and the environment; (3) the atmospheric and stratospheric systems that track wider biogeochemical processes; and (4) the social/economic system in which the activities of humans in transforming less developed into more developed areas impinge upon other humans.

Ecological Knowledge

A desire to know more about ecosystems advanced sharply in the 1960s, prompted by a desire to understand water pollution. At that time the main focus was on the depletion of oxygen in streams as a result of discharge from municipal sewage and industrial works. Hence in these years the standard by which stream pollution was measured emphasized the health of fish populations, and knowledge about the ecological context of streams expanded from fish to aquatic life generally. As time went on, the role of toxic chemicals such as lead, mercury, and synthetic chemicals joined that of bacterial contaminants as a factor to study. Aquatic toxicology as a method of analysis received quite a boost from these scientific interests.

Knowledge about land ecology proceeded more slowly despite the fact that many ideas about ecology had to do with biological succession in the recovery of sites disturbed by farming or logging. The general problem became caught up in the issue of the effect of wood harvest on ecological processes. The main centers of forest research managed by the U.S. Forest Service were not prone to consider this an appropriate subject for intensive research. It was not until other and often competing centers of research—the National Park Service, the Smithsonian Institution, universities or botanical museums and gardens—got involved that much headway proceeded on this issue. When the Society for Conservation Biology was organized in 1987, it marked a new stage in the search for ecological knowledge arising from already well established scientific endeavors in a much wider range of institutions.

Research on plant and animal "wild resources" was ongoing. The earliest pertained to game animals, financed through federally raised funds from taxes on hunting equipment. Gradually, however, the entire range of flora and fauna of ecosystems came under study. The Endangered Species Acts that accumulated over the years provided funds for both population and recovery studies. For animals that migrated over long distances, such as wolves, grizzly bears, and birds, a major breakthrough in knowledge came with the technique of radio telemetry to follow animal movements via implanted or collared radio transmitters. Even the smallest birds and butterflies could be followed over long distances.

Especially significant as a trend in ecological science was the focus on habitat, that is, the environmental circumstances—geological, biological, chemical—within which plants and animals lived. The notion of habitat had firm beginnings in understanding the circumstances of water, food, and shelter required by terrestrial vertebrates such as deer or the habitat requirements of migratory birds along the north-south flyways or the aquatic habitats of game fish. The most dramatic cases involved the habitats required to restore threatened and endangered species, but ecological interest extended to many other species as well. Protecting land and aquatic habitats became a land-use issue and pressed landowners and developers, public and private, to think in terms of habitats as well as individual species.

Ecological knowledge was hampered by the fact that though most ecological understanding involved long-term change in ecological systems, the design of most ecological research was organized around short-term change. Gradually, however, a number of research centers were established that, if supplied with continuous assured funding, could place short-term change in the context of long-term change. The first of these was the U.S. Forest Service research center at Hubbard Brook in New Hampshire. When the debate over acid rain arose, it focused on long-run and short-run interactions, and Hubbard Brook had the most continuous long-run time series of data. Some centers emerged from a ten-year international program that established biosphere reserves primarily for research, and several of these were established in areas relatively undisturbed by human action, such as the Great Smoky Mountains National Park.

Human Health

The search for knowledge about environmental health was heavily bound up with tracking human exposures to contaminants; the relevant science sought to identify exposures and their origin, the relationship to various human diseases, and the pathways through the human body from sources to effects. The sequence from cause to effect aroused considerable controversy through debates over the preciseness of each, the point within the pathway to be measured, or the method of measurement. Yet, amid the continual controversies, research accumulated that provided an ever richer body of knowledge.

All this led to several new perspectives. One was the change in the human subjects of interest. In the 1950s and beyond, the predominant research subject was the healthy adult male worker; most knowledge about exposures to pollution came from occupational health studies. In the 1960s, however, the notion of "community" health expanded the concern from male workers to a concern for people not normally in the workplace: children, women, and the elderly. At first, estimates of the impact of pollutants on the community were extrapolations from knowledge about male adults, but this knowledge, some argued, was distorted because of the "healthy worker syndrome," that is, the fact that most employees selected for work were more healthy than others, hence the effect of exposures on them was much less than on children, women, and the elderly. These groups had to be studied in their own right.

One of the first new groups to be the subject of environmental health science was children. Pediatric scientists worked as individuals or collectively through the American Society of Pediatrics and its Environmental Health Committee; they were joined by scientists from the American Psychological Association who were interested in child development. The first major drive in this direction for environmental science came from a concern about the effect of low-level lead exposure on the neurological development of children. In a similar way, the effect of pesticides in food came to focus on the neurological development of children, and amid considerable controversy, the importance of this problem was enhanced by a study from the National Academy of Sciences in 1992.

Research on the elderly and women came more slowly. Medical interest in the elderly often focused primarily on therapies to cure or repair some aspect of aging, not so much on environmental exposures early in life that might accelerate aging. Cancer was the most studied disease of the elderly, but most of the focus on treating cancer was on reducing deaths for younger people. At the same time, those who sought to emphasize environmental agents as causes of cancer, rather than lifestyle causes such as smoking, were countered on the grounds that cancer was simply due to aging and hence not subject to significant environmental analysis. In environmental science, the emphasis began to shift, slowly but steadily, from efforts to prolong life for the elderly a few years to efforts to reduce developmental problems, thereby enabling people to be healthy and physically able throughout life.

Research on the distinctive effects of environmental hazards on the health of women came even more slowly. Some argued that the health problems of women did not differ sufficiently from those of men to justify the added expense of two sets of environmental health data organized by gender. But other scientists sought to explore health problems particular to women. Some argued that lead absorbed and stored in the bones was released into the body with menopause; others, that chlorinated hydrocarbons such as dioxin or pesticides, which were stored in fat, were present in human breast milk and had detrimental health effects both for women, as a cause for breast cancer, and for children fed on breast milk. Research appeared to suggest that some chemical exposures affecting reproduction were associated not just with female circumstances but also with male sperm, emphasizing that both males and females were affected, but in different ways. And intergenerational research, although still in its infancy, began to identify the close connections between exposures to mothers and effects on the fetus, charted especially in terms of lead, but extending also to some synthetic organic and toxic metal chemicals.

These innovations in environmental health extended the range of diseases that came into focus and tracked chronic low-level exposures and effects. They increased interest in morbidity beyond the more limited efforts to prolong life that had long been the measure of medical achievement. In addition to research on reduced neurological development, there was an increasing focus on reproductive capacity, fetal malformations, and the impairment of immune systems. Declining lung function, which led to respiratory problems for the elderly, received far less attention. Through the interest in reducing such respiratory problems as emphysema, asthma, or cancer, pulmonary science gradually began to identify how biochemical changes in the smaller pockets of the lungs developed slowly from early childhood exposure through adulthood to later effects in the elderly.

Another feature of the new research was the challenge of measurement. The acute effects of high-level exposure were easily followed in such cases as the Donora, Pa., smog of 1948 or the Bhopal, India, isocyanate episode in 1984. But the challenge of modern environmental health was the task of tracking chronic and long-term consequences of low-level exposures. Traditional methods such as epidemiology and toxicology were far more difficult to work out in such cases and bogged down in arguments over other possible causes or the relevance of animal experiments for humans. The evolution of molecular measurement brought a new context into chemical analysis, but the cost of the relevant methods was great and only slowly could laboratories afford to support experiments, let alone large-scale human monitoring.

Environmental hormones—chemicals that mimicked human hormones—became a major focus for environmental health debate in the 1990s because they brought together many of these new developments in environmental science. They shifted the focus of human effects from cancer to the full range of newly emphasized "morbidity" conditions such as reproduction, fetal development, neurological development, or immune systems. The presence of environmental hormones emphasized the role of toxins in their myriad pathways through and around the human and natural environments. It also brought biochemical change in both these environments to the fore as a focus of research. The issue brought together wildlife and humans in one biological context that was far more fundamental than the conventional argument over whether or not animal data could be applied to humans to predict tumors. The endocrine systems of both vertebrate animals and humans were quite similar and had evolved in similar ways over centuries. The power and drama of the environmental hormone issue came from the fact that it crystallized many of the emerging developments in environmental science; it both summarized how tendencies were moving forward and defined new problems and new research to advance knowledge still further.

Biogeochemical Pathways

Environmental science also sought to track the biogeochemical pathways that tended to integrate many aspects of environmental knowledge. This involved a host of scientists in chemical, biological, physical, and geological sciences; hence the term "biogeochemical."

The most significant integrative feature of these elements of environmental science was knowledge about the ways in which biogeochemical agents moved from one place to another, from one medium to another, from sources to effects, and in the process not only migrated but were transformed. The challenge here was to go beyond details of specific chemical reactions to work out the routes that chemicals took in moving through air, water, and land and to identify the changes that took place in them as they moved.

It was far easier to identify changes in the chemistry of streams or the effects of air pollution on plants or the exposures of low-level lead concentrations on child development than to track movement of chemicals through the atmosphere. Much could be understood at one point in the pathway, the beginning or the end, but tracking the process of movement through water or air involved many such points and frequent measurement. It was difficult to identify pathways within the human organism or animals, both domesticated and wild. It was even more difficult to follow pathways from rain through soil to underground water or rivers or through the terrestrial or aquatic food chain. Volatile toxic chemicals such as PCBs presented especially difficult problems because their movement involved a constant sequence of deposition and volatilization that required frequent measurement.

Many chemicals went into the air from varied sources and then migrated to many and varied destinations. In some cases, movement took place in relatively limited or narrow paths, as did radioactive debris from atomic tests in the western United States that then moved in bands around the Earth. Many industrial emissions such as lead or sulfur dioxide were dispersed within the Northern Hemisphere and hence led to arguments that their sources and effects, found upwind and downwind in that geographical area of the globe, were considered to be related. But some chemicals were more difficult to track, such as lead from gasoline-powered motors or DDT, which were widely dispersed within regions and around the Earth.

Knowledge gained from following chemicals through the air developed gradually. In the 1970s DeVinci experiment, air emissions from sources in St. Louis were tracked by balloon as they moved eastward, accumulating knowledge about not only physical movement but also chemical transformations within the cloud. Another study attempted to identify pathways of air deposition in the Northeast by identifying deposited pollutants in terms of their "fingerprint" of combined chemicals and then relating that pattern to emissions distinctive to industries in upwind regions. Tracing pathways of toxic chemicals such as PCBs from the southern Mississippi Valley to Ontario was more complex, since it involved frequent measurements to follow the continuous process of deposition and volatilization, redeposition and revolatilization as the chemical moved along the pathway.

Knowledge about atmospheric dispersion often brought into focus more features of broad atmospheric interconnections. One problem was to measure Arctic smog comparing the "natural" components and the human ones arising from urban-industrial areas of the Northern Hemisphere. Another was to follow the dispersion of chlorine-based chemicals from the lower atmosphere into the upper stratosphere, which led to a reduction in the ozone layer there and permitted higher levels of damaging solar radiation to reach the Earth. These contributions to environmental science were the product of advances in airborne measuring devices.

Water served as the other major medium through which chemicals moved. One pathway involved transport of human and industrial waste via rivers and its effect on biological life, either killing most aquatic life from exposure to toxic and acid mine waste or consuming oxygen as waste biologically degraded. Another was the flow of water from rain, which produced "runoff" from the ground via streets and ended up in sewage discharge. Or the passage of rainfall and dry particulates into soil, transforming soil chemistry through acidification and thus retarding the growth of plants. Still another was the transport of agricultural chemicals, pesticides, and excess fertilizer through runoff into lakes and rivers.

The potential of tracking chemicals through water went far beyond customary research as scientists began to apply mass spectrometry and gas chromatography to fingerprint chemicals so as to describe their movement through water pathways. In one innovative research project, German scientists tracked dioxin discharged in water from a paper mill to the use of that water in manufacturing T-shirts, to the transfer of the dioxin from the T-shirt to human skin, to the bath water of those who wore the T-shirts, to the resulting wastewater in sewage. Such research emphasized the massive gaps in environmental knowledge resulting from the complexity and the cost of the task of measurement.

Economic and Social Development

Knowledge about the human occupation and use of space, one of the most important topics in environmental understanding, advanced far more sporadically than did knowledge about ecosystems, human health, and biogeochemical pathways. The human environmental experience associated with development is one of a continuous decline in the amount of open space. But only a few attempts were made to develop systematic knowledge about this environmental change.

One aspect was the decline in farmland as urban sprawl moved outward and converted farmland to homes, commercial establishments, and factories. The process has been charted somewhat systematically in recent years as a basis for possible action to protect farmland. Far less fully charted is the process by which forest land is often converted to homes, a process that has led to the fact that most eastern woodland is held in "nonindustrial private forests" owned by individuals. Land conversion went on continually, relatively uncharted. For example, the decline in the popularity of summer camps for children, which were usually located on land owned by churches, the Boy Scouts, or the YMCA, often prompted those organizations to sell their land to developers to raise money for other projects.

Land development also hastened the decline of land as habitat for wildlife. As the interest in biodiversity, or the full range of wild flora and fauna, grew in the years after 1970, so also did interest in its habitat. Larger blocks of forest land were carved up into smaller ones by roads, power lines, and homesteads, which increased habitat available to species that can "live with humans" and decreased it for species requiring less disturbance to survive. While aerial photography and historical records permitted some detailed facts, knowledge about these processes proceeded slowly.

Significant change took place in the twentieth century in land beyond the city. Over the years farming had declined and so had the value of farmlands. Often owners did not pay taxes and the land reverted to the state. But after World War II, such lands came to be desired by urban people for their environmental value as recreational lands and locations for vacation and retirement homes. By the 1980s this new demand had far outstripped supply, and the market values of such real estate had risen rapidly. This historic turnaround was well noted by real estate agencies that placed environmental values in the forefront of their land advertisements. Soon the very values that had attracted newcomers to rural areas seemed on the point of being undermined.

Those concerned with these changes attempted to insulate open land from market pressures either by public acquisition or by private land-trust arrangements. In the past, such acquisitions had come from the fact that many lands had been relatively unattractive for development—for example, hilly and mountainous land, wetlands, or infertile lands such as pine barrens. But as time went on, lands held for many years by families as vacation properties continued to have high value for their natural quality, and in many cases these were transferred to land-preservation agencies. Despite the fact that public interest in natural land placed a high value on it, economists were predisposed to associate value with resource development rather than resource protection. Hence only in a few cases were these features of natural values subject to systematic description and analysis.

The most focused effort to acquire such systematic knowledge came in areas around cities subject to intensive growth, where the costs of growth came to be highlighted by those skeptical about it. Growth led to increased public expenditures for roads and highways, sewage works, schools, police, parks and playgrounds, and social services. Studies of the costs as well as the benefits of growth concluded that only with high-value homes did the added returns from property taxes outweigh the resulting costs of development.

One aspect of changes in land use around cities became emphasized in the 1970s amid a concern for rising levels of energy consumption. Suburban development increased energy consumption through commuting from home to work, shopping, or recreational and leisure activities as urban population spread out. Moreover, public utilities such as sewers and water supply became more decentralized and hence more costly in terms of both money and energy. As a result, arguments arose that public policies should encourage people to live in cities rather than to spread into outlying regions. In the late 1970s, the federal Council on Environmental Quality undertook a major effort to document all this in a massive study, *The Costs of Sprawl.* But such efforts to acquire knowledge about the environmental workings of development were rarely undertaken.

SCIENTIFIC CONTROVERSY

The evolution of environmental science involved intense controversy in almost every facet of science. There were disputes over how scarce scientific resources should be allocated. Amid the prevailing bias toward the physical sciences, federal funding for ecological research experienced considerable disadvantage. And those who feared the regulatory implications of new knowledge often objected to scientific work that might give rise to that knowledge. The demise of the Ecological Research Service established by the U.S. Fish and Wildlife Service in the late 1960s or the intense opposition in the mid-1990s to the ecological work of the National

Biological Survey can be attributed to this attitude. Several evolving realms of ecological knowledge—regarding wetlands and endangered species on the one hand, and climate change on the other, with its implications concerning stratospheric ozone and global warming—met with considerable hostility from the industries that could be affected by such knowledge.

Such controversies, however, often paled in the face of controversies over the implications of scientific research: what did it all add up to? The direct relevance of this question to policy and action meant that the media, policymakers, and scientists were all preoccupied with the meaning of the science. The language of dispute often took the form of buzz words such as "flawed science," a relatively polite term; "good science and bad science," as somewhat more dramatic terms; or "junk science," in more extreme language. Such words as these flowed continually as issue followed issue in the meaning of science, whether it dealt with human health, ecology, biogeochemical pathways, or development. It is tempting to evaluate the assessment of each issue separately. Yet it is also possible to describe the overall patterns of such controversies as they added up over the years.

The evolution of environmental scientific knowledge involved a continuum from the earliest stages of the presentation of ideas and initial evidence through successive stages leading toward some relatively firm conclusions. Construct a band from right to left across a spectrum and mark out stages in this spectrum as knowledge develops. Although the steps along the way can be described in terms of the knowledge itself, it is more important to think of the process as one of scientists discovering, debating, arguing, disagreeing, and then in many cases developing a degree of agreement while those who continue to disagree become peripheral to the scientific debate and seek nonscientific political allies to continue their cause. "Conclusive proof" is often called for by that minority, but it does not occur because its requirements are beyond the reach of science.

This process begins not with hypothetical questions, which are rarely taken seriously, but with empirical studies or some unusual reasoning from existing studies or some attempt to make new meaning out of quite diffuse research. Think of this as the segment of the band in our continuum in the 10 percent range from the right end. At this point other researchers become interested and the focus shifts from one statement to confirmation of the statement by others; perhaps earlier studies are replicated or another method is used to get at the same subject or the research is applied to other subjects. This takes place between the 10 percent and 20 percent stages on the continuum from the right end of the spectrum toward the left.

The debate now enters a new stage in which sufficient knowledge has accumulated to arouse vocal skepticism. Enough critics have come onto the scene to begin a vigorous debate with advocates of the new knowledge over the interpretation of the facts. That they do so reflects an important stage in the evolution of science because it means that the science is convincing enough to a sufficient number of scientists that it cannot be ignored, hence the stage of vigorous vocal opposition.

The third stage in the band of the spectrum, at the 20 percent to 50 percent level, is a stage of vigorous opposition that involves the organization of research designed explicitly to challenge the new knowledge. Some simply seek to repeat work already done to disprove it. Others insist on the application of a new method or new subjects to reduce its application or to argue that what was thought to be "typical" is, in fact, not so. At this level, the issue of research design comes into play as a major factor in scientific dispute. If one wishes to question research, one simply devises a different research design. In fact, many of the disputes among scientists center around the relevance of one research design versus another; the resulting arguments, however, are so esoteric to the media and the public that this realm of scientific interchange rarely surfaces in a public way, even though it shapes much of the discourse about the meaning of scientific knowledge.

The debate now continues with vigor and intensity, frequently taking on an intensely personal tone between scientists. Different scientists will often see the whole question through the narrow vision of their own methods or the ideas they acquired in their early training, which they continue to apply. Others will follow the interests of those by whom they are employed, framing the issues so as to provide the best result for their employers. Still others will be drawn into the mélange of debate by the dispositions of personality, which make some feel comfortable questioning older scientific views in the face of the new while others do not without a much higher level of proof. It is at this stage, halfway through the spectrum and above, that one confronts the way in which science, by virtue of its enormous range of specialized knowledge and method, generates its own internal disputes, often takes on the form of personal animosities among scientists, and creates social and psychological barriers to movement across the spectrum of knowledge development. It is here that the language of debate takes on the form of "good science" and "bad science" and fosters the use of these epithets as a substitute for the substantive issues at stake.

Finally in the evolution of scientific knowledge, the issue at times, though rarely, reaches the 75 percent level and above. Demand for "conclusive proof" becomes commonplace even though that level of knowledge seems impossible to achieve, since the meaning of science at each stage in the process involves considerable personal judgment. Migration of scientific activity across the spectrum involves "levels of proof" that one demands as a basis for drawing a conclusion. Its culmination is a stage of "workable agreement" rather than "conclusive proof."

To the public, the question is usually one of "is there enough proof," an issue that the media takes up in order to bring simplicity out of complexity. By the same token, the language of debate on all sides veers toward simplicity for purposes of communication with decision makers, fellow professionals, and the public. This complexity of debate, arising out of the complexity of arguments over proof, generates opportunities for those who wish to slow up application of scientific knowledge and establishes a context of caution on the part of decision makers in public agencies. Their watchword is "insufficient proof."

Perhaps the most compelling of all diversions in the task of assessing environmental science is the frequent assumption that one can make firm judgments as to the state of the environmental world. New knowledge about the environmental world proceeds slowly, and with each step in the process, more ignorance than firm knowledge is achieved. But the pressure for firm conclusions tempts one to express considerable confidence about what that unknown is all about. Expressions of such confidence reached their peak in the debate over comparative risk assessment. Demands for such "conclusive" assessment carry with them the assumption that the problem is a simple one of providing experts with the authority to make decisions about the relative rank of risks based on scientific knowledge. Yet as the debate unfolds, it becomes increasingly clear that the relevant scientific knowledge is not available and may not ever be available as the parties to dispute demand, and perhaps even that science is incapable of developing relatively firm knowledge about a subject as complex as the environmental world.

CHARTING THE ENVIRONMENTAL WORLD THROUGH SCIENCE

The various scientific initiatives from exploratory, defensive, and managerial science served as powerful impulses to advance environmental knowledge. The disputes over levels of proof tended to retard those advances, as those who required more proof were able to exercise a powerful influence in both undermining conclusions and slowing up the application of knowledge. Yet, if one compares the state of environmental knowledge in the year 2000 with the year 1960, one can only conclude that the historical evolution was persistent, profound, and remarkable as a stage in the history of science. If one is absorbed in the episodic disputes over the direction, the methods, and the meaning of environmental science, one despairs of making sense of those disputes. One sees only a sequence of truth and error, of bad science and good science, of rationality and emotion. But this hardly describes the real world of scientific exploration, its impulse, its direction, and its results. At the end of the century, far more was known about the environmental world than was the case forty years before; the emergence of environmental interest on the part of the public at large and the world of scientific endeavor was largely responsible for that change. Despite the continual resistance to exploratory science on the part of defensive science, the environmental world has persistently undergone new inquiry, new exploratory ventures, and newly charted landmarks.

Exploratory science's contribution to this historical evolution was one of working on the frontiers, of being drawn on by the intriguing possibilities of turning the unknown into the known. At the same time, exploratory science was quite controversial just because it did work at the frontiers, using new methods to get at difficult problems of knowing, following up the implications of past research, and seeking to make the results acceptable. How the environmental world works in ecology, environmental health, biogeochemical pathways, and human health and development was discovered and confirmed by exploratory science.

Defensive science made quite a different contribution to this development. By demanding further investigation and higher levels of proof, it fostered the task of ever-more-demanding confirmation of exploratory science rather than led the way to new paths of knowledge. Much of that effort, at the same time, lay in seeking to divert attention from cause-effect processes inherent in scientific understanding to those particular features of knowledge that bore especially on the regulated industries. The accumulation of defensive knowledge, therefore, was negative rather than positive, emphasizing errors and unknowns in knowledge already accumulated rather than the continued accumulation of knowledge at the frontier.

Managerial science, in still a different role, sought to assess the meaning of knowledge so as to establish a workable consensus on which to base action. This led to its dual task of assessing existing knowledge and fostering exploratory science guided largely by "gaps" in knowledge peculiar to a particular proposed action rather than by gaps defined through the task of more wide-ranging scientific exploration. Managerial science was decision driven; requirements for action resulted from the need not just to advance science but to encourage decision making that would be relatively immune to challenge from those affected by those decisions. Consequently, the meaning of science was heavily shaped by law and the lawyers who sought scientific justification for their arguments on behalf of their clients.

Amid the interplay of exploratory, defensive, and managerial science and the involvement of these three types of science in the world of political strategy, it would seem rather remarkable that one could speak of a cumulative body of environmental knowledge at all. Yet that accumulation did take place, and by examining its stage-by-stage evolution, one can chart the evolution of environmental science to a firm position in the history of science. Environmental science is one of the most vibrant, vigorous, and extensive advances in science in the last half of the twentieth century. With this perspective in mind, the most intriguing question is the degree to which the cumulative knowledge of environmental science has established an objective context relatively free from continual political dispute. Has that context exercised some discipline over those debates to generate more independent understanding of the way the environmental world works?

12 THE ENVIRONMENTAL ECONOMY

If there is any issue that continually shapes the debate over environmental affairs it is that of the "environment and the economy." The links between what traditionally has been thought of as the economy and what is now thought of as the environment are closely interconnected. Every facet of long-standing ideas about the economy has an environmental dimension. Environmental affairs have brought a new perspective to bear on traditional activities such as production and income, jobs and investments, consumption and markets, capital formation and maintenance, resource use and depletion, and the distribution of benefits. Environmental issues have generated a wide range of tensions between older economic practices that arose with little thought about their environmental implications and newer ideas influenced by the environmental consequences of traditional economic activities.

One can summarize the environmental facets of economic activity as the emergence of bits and pieces of an environmental economy. To do so emphasizes at the outset major distinctions between the old and the new and defines some of the fundamental tensions between the two. First, the environmental economy is defined not so much by factors of production as by patterns of consumption. Economic activity has long been described as two sides of the same coin, production and consumption. Some economists—for example, Adam Smith in *The Wealth of Nations*—have argued that production depends on consumption and that without the markets by which consumption becomes activated, production would be of little significance. At the same time, however, producers have shaped most of the thought, most of the data and information, and most of the public policies pertaining to economic affairs. Hence historians reconstruct economic history based on a wide range of historical data about production but have relatively little data about consumption. The environmental economy played its initial economic role in representing new types and standards of consumption.

Environmental affairs pertain to what people want in their lives, what kind of circumstances they desire surrounding home, work, and play. These wants and needs constitute markets that stress experiences and surrounding circumstances, and hence have come to be described by the term "quality of life," a qualitative dimension that adds meaning to the traditional term "standard of living." As such they are part of a wide range of qualitative conditions, each constituting a distinctive market, that have come to define new economic sectors—the health economy, the recreational economy, the learning economy, the amenity economy—which shape production to respond to new facets of consumption. The previous wants and needs have not gone away, but they have been modified and added to by environmental values. It is among these "market forces" that we find the significance of the environmental economy.

THE ENVIRONMENTAL ECONOMY: BURDEN OR OPPORTUNITY?

The emergence of the environmental economy has led to two different views about its economic role: (1) that it is a burden, a cost, and (2) that is it an opportunity, a benefit, similar to material benefits like houses, appliances, and automobiles. For those accustomed to older ways of thinking, it is a burden on the more important economy of material production. It is the older economy that is fundamental, that alone produces a "surplus" from which environmental consumption can be financed. From such a viewpoint, environmental consumption is a luxury. This distinctive burden placed on environmental affairs is itself an important feature of environmental politics. It arises in almost every environmental initiative—"can we afford the cost?" is the question asked, and it overshadows the obvious potential of environmental market opportunities. Hence it has remained for those involved in these market opportunities to press their case in selective niches of economic thought and action.

The aesthetic and appreciative consumption of nature is one of the more widely emphasized facets of the environmental economy. The human experience of nature, increasingly prized by many people, gives rise to considerable personal expenditure. Some involves the purchase of goods to enable one to enjoy nature, such as binoculars or photographic equipment or birdseed. There are expenditures to travel to areas to enjoy: parks and forested areas, coastal areas, as well as foreign countries in what became known as "ecotourism." One aspect of the nature economy was the search for environmental knowledge, an integral part of the evolving learning economy that encompassed knowledge ranging from guidebooks to the identification of plants and animals to magazines about natural history or scientific books about the way the natural world worked. Nature programs on television became popular, and the well-stocked shelves of bookstores testified to the widespread interest in "nature" publications.

From time to time there have been efforts to calculate the role of these activities in the economy, and the most extensive of these have been carried out by the U.S. Fish and Wildlife Service. Every five years this agency identifies the number of people engaged in hunting, fishing, and appreciative wildlife activities and the amount of money spent on them. The debate over public forest policy and especially over the newer environmental and ecological uses of the forest led the U.S. Forest Service to calculate their economic implications for the national forests. By 1997 the increase in recreational income prompted the new Chief of the Forest Service, Mike Dombeck, to announce that by the year 2000 income from outdoor recreation would be thirty times greater for the Forest Service than income from resource extraction, including wood production.

This more passive role of the environment in the economy wormed its way only slowly into economic thinking. Its major inception took place in the Northern Rockies and the Pacific Northwest and had its roots in examining the economic base of the changing northern and western economies. The leading figure in these ideas was Thomas Power, chair of the economics department at the University of Montana in Missoula. Power's initial research identified the close connection between population growth in western Montana and the existence of areas incorporated into the National Wilderness System; rather than constituting an economic burden, wilderness was an opportunity to add a significant component to the regional economy.

Studies of trends in the economies surrounding Yellowstone National Park and the six national forests around it marked rapid changes from the old to the new economy. Within the national forests themselves, the extractive economy provided less than 5 percent of all jobs. These studies were augmented by accounts of similar changes in Oregon, where, despite the public image that the wood-production industry was the state's most important job sector, data brought forth by an Oregon State University economist ranked it sixth in employment and charted its steady decline over the previous thirty years. By the early part of 1997 the thinking of economists in the Northwest had changed to the extent that Power was able to persuade over seventy economists in the region to support a statement that environmental quality was one of the major reasons for the strength of the region's economy.

The role of nature in the environmental economy had many and varied dimensions. One was the way in which several communities throughout the country had organized nature-based activities to attract visitors; in several such cases the older extractive-based economy of the community had collapsed and a newer environmental economy had replaced it. Such was the case with Ely, Minnesota, which began to prosper from the establishment of the International Wolf Center; or with

Dubois, Wyoming, which replaced its lumber mill with a Bighorn Sheep Museum and used it as a magnet to attract artists, tourists, and residents. The National Raptor Center in southwestern Idaho, by the sheer force of the economic strength of its museum, research, and tourism, overcame the resistance of the far less lucrative ranching economy, which had unsuccessfully resisted the innovation.

Many of these ventures underline the importance of nature as an attraction for locating homes. This has varied facets, one of which is the way modern communications make it possible for people to live in one place and do business in another. This gives rise to "footloose" industries, of people who live in the towns and smaller cities of the Rocky Mountain Region and maintain their business elsewhere. Another facet was the role of attractive natural settings as retirement communities, which draw older people. Economists could demonstrate that the economic contribution of this population to the community from their retirement income and investments was much greater than the income from extractive economies. Still another aspect was the attraction of open space and natural lands as residential areas in which other and older forms of production were maintained. Proximity to natural-environment-based recreation was cited positively in many reports about factors influencing the choices of business executives, academics, scientific and technical professionals in locating their desired place of employment.

Extensive financial support for land conservancies, which grew steadily over the last decades of the twentieth century, reflected a consumer interest in open space as a desirable part of the community environment. Membership in land conservancies grew, and fund-raising for individual conservation projects attracted considerable support. Land conservancies became important recipients of memorial gifts to honor those who died, emphasizing the degree to which conserved land was thought of not just as a momentary expenditure but, as some economists have designated it, for its "bequest value" as well. Land-conservancy expenditures play much the same role as do fiduciary trusts, involving a degree of permanence over several generations that provides a continuing capital base for a valuable community asset.

Environmental consumption has many facets. One is the market for consumer environmental products, such as organic foods. Each year the popularity of organic foods grows. First came specialized organic food stores, and then more conventional food markets added sections for organic foods. These advanced more rapidly in the west and on the east coast and proceeded far more slowly in the south and the central states, especially in rural areas. Organic foods represent considerable consumer buying power, and they also demonstrate clear and well-recognized differences in life-style between the old and the new. Organic foods were resisted by the agricultural products and agrochemical industries on the grounds that they were detrimental to the food economy; these industries were not willing merely to let this new market develop according to consumer demand, but went out of their way to attack its economic soundness and even its legitimacy. But organic foods entered the economy more and more each decade as consumer demand for them increased.

Equally significant was the market for pollution control and pollution remediation technology. This included technology to reduce pollution levels all along the production chain, to clean up degraded areas, and to identify and monitor the transport and fate of environmentally damaging chemicals moving through the air, water, and land. All this gave rise to new environmental industries. Industries that manufactured equipment to remove air-polluting chemicals at the point of emissions were organized in the trade association Institute of Clean Air Companies. Some industries manufactured water-pollution control technologies both for municipal sewage plants, as they advanced from primary to secondary to tertiary treatment, and for industries that were required to reduce their effluent. Running throughout all of these pollution-control programs was the need to identify concentrations of chemicals, to track emissions and discharges both as they left the source and as they migrated in the atmosphere, waters, land, and in food chains to affect the wider environment. Such opportunities created a vibrant monitoring industry.

Legislated environmental standards, such as new or improved standards of ambient air quality, established additional new markets for the industries providing the equipment to meet those standards. Economists and institutional leaders often looked upon those standards as burdens on existing industry, but they were in fact market opportunities. Because the political influence in such matters was exercised far more by those defending existing industry than by those fostering new technologies, the role of market innovation was often overlooked. Representatives of the emerging environmental industries often sought to advocate advances in standards as a means of promoting new environmental industries, but they were not able to influence dominant economic thinking.

In the mid-1970s considerable interest arose in new markets for those who would supply more energy-efficient products, including energy efficiency in manufacturing, in the operation and management of industrial plants and commercial establishments, in home heating and lighting, and in appliances. A host of new analyses of energy use and waste came to the fore. Legislative proposals to require more fuel-efficient cars and trucks and more fuel-efficient household appliances were introduced. In both cases, conventional industries that were less energy efficient vigorously opposed the innovative standards. But in some areas, especially in household appliances, some manufacturers took up the market opportunities and wormed their way into the economy. Automobile manufacturers made some changes that raised the mileage performance of many cars, but overall they staged a successful resistance to the standards.

This new production was popularly known as "green production," and from time to time its larger economic significance was recognized. This was especially true after the World Environmental Congress at Rio de Janeiro in 1990, when a number of economic leaders in the United States became enthusiastic about the potential for selling American environmental goods and services abroad. Some striking examples were already on record, such as when manufacturers of automobile catalytic converters, required in the United States, were primed to enter the European market when they were required there as well. After the Rio conference, several cities in the United States sought to capitalize on this momentum by holding conventions touting their regional environmental industries and the benefits to their cities from the new international market opportunities. At the same time several new publications took up the advocacy of "green industry," such as the magazine *In Business,* which became a major source of information in tracking its growth and development.

ENVIRONMENTAL ACCOUNTING

Various connections between environmental objectives and more traditional and conventional economic activities kept coming to the fore largely because the "economic" implications of environmental affairs could not be ignored. Those implications had both positive and negative consequences; hence those who held more traditional beliefs about the relationship were forced, in some way, to take them into account. The connection could be observed on two different levels: one, economic thinking of institutional leaders in government, the professions, and think tanks; and the other, in that of leaders in private economic enterprise. The two were not, of course, unrelated, but their responses to the persistent interconnections stemmed from different origins and each had its own dynamics. Both groups of leaders, however, thought of the connection in terms of traditional elements of economic activity: jobs, investment, and profitable enterprise. And both, in turn, presented arguments within the traditional context of national economic accounting.

The focus on jobs played a significant role in the Clean Air Act of 1970, which included a provision that the EPA should issue quarterly reports on its investigation of firms that had allegedly gone out of business because of pollution-control requirements. The issue was whether or not closings were due to technical and market obsolescence or pollution controls. Many a firm that was lagging in market efficiency sought to blame its problems on the new requirements, and the task was to sort out the two. Both employer and employee organizations championed the reports as a way to pinpoint the job losses to the new air-quality program. But the ensuing reports demonstrated that job loss was due to production obsolescence far more than to pollution-control requirements. Both labor and industry lost interest in them, and they no longer became a factor in administering the Clean Air Act.

At each stage of environmental debate, industry and labor took the lead in warning about the adverse impact of some new environmental policy on the "economy." Periodically, economists took up the issue to try to determine the accuracy of such claims. The more reliable and comprehensive studies reported that environmental programs had a net economic benefit, that there were some adverse effects on employment in existing firms but that new economic ventures stimulated by environmental programs offset the negatives. Environmental activity had given rise to considerable private enterprise, and at times the positive benefits of environmental programs had helped to offset weaknesses in other sectors of the economy. But though such views contributed momentarily to the debates, they had little staying power in the face of the argument from economists and industry that the impact was overwhelmingly negative.

Most economists accepted the idea that pollution created an economic burden to the society it affected rather than a benefit; it was what they called an "externality," that is, it involved a cost not internal to the firm's own balance sheet, but a cost imposed on society. Therefore, they argued, "external costs" should be "internalized." In some fashion, firms should be required to incorporate them as costs into their internal balance sheets, and this, it was usually suggested, should be done through a tax on each firm in proportion to the amount of its pollution. But though the pollution tax was well accepted by the environmental community, it was rejected by the polluting industries. Hence, although the idea of "external costs" appeared in the professional literature in the 1970s, and the pollution tax was a central idea in one of the most widely accepted publications about "market forces" in the 1980s, it succumbed to this political resistance on the part of sources of pollution and failed to become a part of serious national debate.

Economists were equally enthusiastic but in the long run equally unsuccessful in drawing attention to the costs imposed on the economy by government subsidies to industries that were environmentally harmful. The environmental community persistently argued that many aspects of government spending promoted economic development that had undesirable environmental consequences. Federal agencies constructed large dams that were economically inefficient and destroyed the natural regimen of rivers. They drained wetlands that were more valuable as undeveloped flood-control areas. They promoted the construction of homes and development in environmentally sensitive areas such as floodplains and coastal areas, and then government agencies were called upon to provide disaster relief when the inevitable came. Very early on in the environmental area, some economists challenged the economic analyses used by the U.S. Army Corps of Engineers in their large-scale waterways construction programs, and environmentalists took up their arguments, which helped to reduce appropriations for such projects.

The challenge in most economic analysis was that the costs of these programs were easier to quantify than the benefits, which were far more difficult to assess and therefore received far less attention. This was especially the case with environmental benefits that were not traded in the market and for which market prices were not available as a measure of value. One could take a stab at this measurement by identifying benefits in terms of the cost of repair. How much would it cost to clean up a polluted stream or repaint a house damaged by air pollution? But how to measure the way in which air pollution created discomfort or reduced the enjoyment of daily life? How much to value the natural environment as an object of enjoyment or learning?

Some economists, intrigued with the problem of how to place a value on such matters, asked people in surveys how much they would pay for a given amount of environmental improvement, a technique known as "willingness to pay" or, in the language of economists, "contingent evaluation." Pressure to find some way to measure a wider range of environmental benefits gave contingent evaluation some momentum. Much of the drive came from programs to clean up aquatic systems damaged by hazardous-waste dumping or spills. How to measure the value of an aquatic system beyond simply the commercial value of the fish destroyed? Federal agencies took up contingent evaluation and the courts looked upon it with favor. But progress in using it went slowly because of the continued opposition from the regulated community, which felt that it would only add to the costs that industry would have to bear for environmental remediation.

Over the years, controversy about these aspects of the relationship between the environment and the economy came to rest heavily on the issue of relative costs and benefits of environmental policies. The problem of calculating the balance between the two was pressed continually by the environmental opposition, who argued that many environmental programs had been adopted without taking into account their costs in comparison with their benefits. This issue of relative costs and benefits had been an integral part of the formation of environmental policy from the beginnings of the environmental era. Trends in that debate, however, had given rise to increasing complexity in the proposed analyses. The environmental opposition would argue that some costs had not been sufficiently taken into account and that these additional cost factors should be incorporated into the analysis. In response, then, the environmental side of the debate would develop strategies to identify the environmental benefits more fully as a means to establish some counterweight to the active role of cost analysis. They also predicted lower costs, which often were closer to the mark.

As the thrust and counterthrust of cost-benefit analysis proceeded, it became increasingly complex and idealized, cast in the argument that cost-benefit analysis had really never been used and that now it should, finally, be used. By the 1980s participants in the debate had been forced to accept the fact that little information required by idealized cost-benefit analyses was available. Environmental science had given rise not to firm knowledge but rather to a sharp understanding of how little environmental knowledge was available to identify either benefits or damage. In 1989, the EPA Science Advisory Board identified a fourfold range of environmental benefits that should be brought into such analyses: mortality, morbidity, ecological, and welfare, establishing a benchmark of what should be known in order to conduct acceptable cost-benefit analyses. Because little was known about each of these divergent sets of problems, it brought to the whole debate an overlay of ignorance and uncertainty that constantly shaped its direction.

Amid this emerging complexity and ignorance a political drive arose in which industry, most economists, and many institutional leaders joined in a simple strategy: use cost-benefit analysis in making decisions under the guidance of "sound science." The drive achieved considerable headway in Congress and moved forward despite the wide acknowledgment that limited scientific knowledge was available for its professed method of making decisions. Because of the well-understood ignorance about such affairs, the proposal seemed to rely more on mechanisms for making decisions in the face of ignorance and uncertainty. Some argued that the decisions should be taken entirely out of the customary realm of legislative and judicial institutions and given to an executive body that had ultimate authority. The congressional proposals, however, continued to give the EPA authority to make cost-benefit decisions, but they also made it easier to challenge EPA decisions in the courts.

While this debate was proceeding in Congress, a quite different debate was going on in the states, shaped by EPA funds that encouraged each state to take up the issues identified by the 1989 report of the Science Advisory Board. Here state "comparative risk" processes moved forward as a basis for making environmental choices. They proceeded in two directions: first, a scientific advisory body that assessed the adequacy of the science to make comparative judgments about risks, costs, and benefits, and second, a public advisory body that ranked "relative environmental risks" facing the state. These proceedings were far less ideological and partisan than those at the federal level, and among the states two general conclusions emerged: first, that the scientific data for an idealized comprehensive comparative risk process was not available and probably never would be, and second, that the relative rankings of risk were essentially value choices that should be made not by experts but by a public process to make sure that they were sufficiently acceptable to the public at large.

MARKET FORCES AND PUBLIC REGULATION

As environmental programs progressed, the environmental opposition developed the argument that environmental regulation was ineffective and that "market forces," that is, private enterprise, would produce better results. Although the argument was applied by free-market theorists across the board in objecting to all environmental programs and regulations, it was taken up primarily by those affected by pollution controls. In earlier years, industrial sources of pollution resisted efforts to clean up; now they were in a far more receptive frame of mind and could be trusted to take up effective pollution-control efforts on their own. Environmental responsibility, it was argued, was now inherent in effective business competition. Hence the regulatory system should be relaxed and greater emphasis placed on market forces to do the job. "Market forces" became the slogan of the day, a slogan that spearheaded the drive to relax regulation. As in many areas of environmental affairs, sound bites emerged to set the terms of public understanding, and in this case those terms were statements that market forces were now being harnessed to advance environmental objectives.

The claims to accuracy in this argument are more than dubious, for it was difficult to detect any significant changes over the years in the strategies of the regulated community and its degree of commitment to environmental improvement. There was some reason to believe that the regulated community had become reconciled to some regulations required by law; yet the continued demand for more "business-friendly" regulation and the continued task of enforcing the law seemed to weaken that conclusion. Much of the long-standing problem in environmental affairs had been the resistance to accepting regulatory requirements, and that continued to be the case. At the most, therefore, conclusions about the degree to which the programs had been accepted were problematical.

What could be said with considerable confidence is that over the years the standards for environmental improvement, in contrast with the techniques for achieving them, had come not from the regulated community but from the public at large. And they continued to do so. At almost every point in the history of regulation the regulated industry had fought against standards that specified the goals of environmental improvement. The water-quality standards of making all streams "fishable and swimmable" came not from the sources of pollution but from the affected public. The drive to establish ambient air-quality standards for lead and to eliminate lead from gasoline came not from the industry but from the public health sector. It was industry that led the drive to relax standards for ozone in 1978, the only case in which an air-quality goal was scaled back, and then in 1996–1997 fought vigorously against its reinstatement to the original 0.08 parts per million. The effort to make the regulatory system more "market based" or "user friendly" seemed to involve the desire to relax rather than improve environmental quality goals.

The Clean Air Act of 1990 was touted as a brilliant example of the effect of market forces in air-quality regulation. Yet the major feature of that law, that sulfur dioxide emissions be reduced by 50 percent, and the allocation of that reduction to specific power plants was a product not of market forces but of a legislative mandate. The utilities required to reduce emissions were rewarded with "emission credits," which they could sell to others who might want to buy them; the credits were a subsidy in the amount of the market value of the credits. Added to those arrangements was the program to reward utilities for cutting emissions even further below the required 50 percent by giving them equally salable credits. The major market force in the program was that the mandated reduction of sulfur dioxide emissions created markets for private entrepreneurs who could help meet those requirements by providing low-sulfur coal or scrubbers.

The tenuous nature of the role of market forces was displayed in some of the new strategies promoted by the Clinton administration to bring greater freedom to entrepreneurs in the air-quality program. The general idea was that those who demonstrated a commitment in practice to reduced emissions would be given greater freedom from the regulatory requirements. But such strategies also carried with them the obligation on the part of the industrial sources to establish a higher level of environmental performance in terms of a lower level of emissions. This proved to be the fly in the ointment in the so-called XL program fostered by the administration. The few industries that agreed to take up the XL program wanted to enjoy greater flexibility in how they achieved the lower emission goals, but they simultaneously resisted higher standards. Rather than demonstrate the value of market forces, the XL program merely pointed out more precisely just how limited reliance on market forces could be.

The most profound relationship between environmental regulation and the market, however, came from the way in which the regulatory system provided a new context for competition. One effect of this was that business firms often sought to locate in states and localities with weaker environmental cultures and weaker environmental standards and enforcement. Such places were well known to the regulated industries. These factors, however, were not uppermost in the location decisions of firms; they were touted more frequently in the public relations of businesses and governments rather than in their behavior. A second effect was the way in which the larger firms were able to enhance their competitive position vis-à-vis smaller firms by being more alert to environmental requirements and by having greater legal and technical resources to facilitate their response. Thus in programs to clean up older, leaky underground storage tanks at gasoline stations, larger firms could finance the shift to a more modern technology and with their superior legal resources even take advantage of public programs to help finance the change. Smaller, "mom and pop" service stations were less alert to the requirements and less financially able to comply with them. And still a third outgrowth was the intense competition that emerged when multiple sources were clearly responsible for an environmental problem. Firms in one line of production, and thus one source of the problem, placed blame on others in the competition to reduce the impact of regulation. Thus in the case of ozone in the lower atmosphere, the utilities and the complex of motorized vehicle interests blamed each other for the cause and hence sought to shift the burden of cleanup to the other.

Competition between states and cities for new economic enterprise became an even more powerful instrument in the attempt by industries to shape the regulatory system to their advantage. Cities and states did not appeal competitively to business enterprise on the grounds that they had a more vigorous environmental program and maintained a higher level of environmental quality but instead on the grounds that they maintained an environmental program more friendly to business. In the early stages of environmental protection, a provision in the federal law allowed states to maintain air-quality standards more restrictive than required by the federal program, and six to eight states did so. But over the years the industries tackled each of the relevant state legislatures and successfully brought to each state the requirement that its standards be no "tougher" than the federal standards.

These efforts by industry to establish a "level playing field" defined in production terms rather than environmental ones were warmly welcomed by state and

urban governments, which had intensified their efforts in the environmental era to secure more industry, employment, and income. The dominant watchword was "growth," and the only environmental requirement that was brought to bear on this strategy was whether or not the state or the metropolitan region was in "federal compliance." While those associated with advancing environmental quality stressed advances in environmental knowledge and the importance of environmental values to the state and the community, the dominant views of both state and municipal leaders tended to stress the necessity to abide by requirements emanating from the federal government.

ENVIRONMENTAL EQUITY

Economic equity has long been a major feature of economic analysis, and the equity of public policies has long been a feature of policy analysis. So it was not surprising that environmental equity should emerge as an important issue in the environmental economy. Much of this feature of environmental affairs emerged only on an issue-by-issue basis, cast in terms of the "fairness" of each environmental program or of equal access to the decision-making process. But rarely did it come to be codified or generalized in the same way that equity, or the more popular term "justice," did. In the 1980s, however, environmental equity came to be associated with the issue of local siting of potentially harmful industrial enterprises and, more precisely, with their siting in black and Hispanic communities or in lower-income communities that had limited political power to resist. This led to the development of the issue of "environmental justice," focusing especially on racial and cultural inequity, but it did not bring into focus the larger issues of environmental equity that had long been involved in environmental affairs but rarely described in such terms.

When natural-resource issues were dominated in the nineteenth century by the distribution of public lands and resources, equity issues involved choices between the distribution of resources to homesteaders and small-scale enterprises on the one hand or to larger enterprises on the other. Some of the support for the transition from distribution of public resources to private parties to permanent public ownership and management came from those who sought to end the ability of those entrepreneurs with greater capital resources to unduly influence public policy and thereby enjoy wider use of public resources. Such efforts were not always successful. Despite provisions in the federal reclamation law, for example, that limited recipients of federally financed water to owners of 160 acres or less, large landowners secured a disproportionate share of the water. Leasing of grazing land under the Taylor Grazing Act and the Bureau of Land Management, although touted as a program for the smaller rancher, was highly skewed so that ranchers with larger holdings dominated the program.

Much traditional resource management and many newer environmental programs, however, involved a wider distribution of benefits. Recreation and fish and wildlife activities on the public lands benefited people of widely diverse means. Those who came from farther away to enjoy the public lands tended to be of upper-middle incomes, whereas those who lived nearby—for example, day users of wilderness areas, who were the largest group of such users—averaged lower levels of income. Hunting and fishing could be enjoyed by users with a wide spectrum of incomes; hunting licenses were perhaps the smallest portion of the cost of such activities, far less than the equipment used, and the right to "take" an animal or fish was well within the means of the average cost of recreation for people with modest incomes.

Environmental air- and water-quality programs carried an even more significant element of environmental equity, since they involved a widely shared environment. For the most part, air and water were not carved up into private ownership such that more could be owned by some and less by others. Hence when programs arose to reduce waterborne disease by purifying drinking water or to improve air quality by restricting emissions from industries or automobiles, they benefited the entire community. Thus, they were first administered by a U.S. Public Health Service division known as a "community health" bureau. Because pollutants migrated through air and water to affect people in many regions, efforts to reduce their impact involved the equitable treatment of individuals over a wide area.

Land-based pollution was a different matter, for it affected the communities where the industrial sources or the waste sites were located. Since no community wished to be subject to the pollution from these sources, the locating industry had the problem of either obtaining the consent of the community or being able to override its objections. Some communities were persuaded to accept the environmental damage by financial contributions to their towns, but many were not and resisted the location of polluting industries in their areas. Inequity often lay in the implicit political capabilities of the community. If it were more affluent and more politically active, the siting industry simply avoided it, choosing instead to tackle communities that had fewer financial and political resources.

Pollution programs involved a wide range of equity issues in relation to different genders, different ages, those of different genetic inheritance, or those of different susceptibility to disease. In the 1950s and 1960s, most attention about environmental health was paid to male employees, but subsequent environmental issues brought into the public debate the effect of exposures on the elderly, women, and children. This came about slowly, with each stage involving a new phase in the general debate over equity. One argument, for example, was that the elderly should not play as prominent a role in the statistics about environmental harm as other populations because their health problems were due to aging rather than environmental exposure. It was only gradually accepted that the health of the elderly did not involve simply living longer but also enjoying a high quality of life through good health after retirement. Another argument was that the health problems of children were no different from the health problems of adults. It was only gradually accepted that child development involved its own distinctive physiological circumstances and that exposures of little consequence to the adult had far more serious effects on the developing child. Moreover, traditional beliefs that the birth defects of children were due to influences from the mother were challenged when it was found that defects in the male sperm due to toxic exposures at work could also have effects on the fetus. Only gradually did physicians recognize that the placenta afforded no barrier to the passage of toxic substances or disease organisms.

One could readily argue that to be equitable, programs designed to control exposures to humans should take into account all of those potentially exposed and treat them equally. But those who stressed the cost of pollution-control policies continually pressed the idea that some people were more important than others and hence to be "cost-effective," policies should protect only those of highest priority. Any statistical distribution of either exposure levels or human effects involved a curve of "average" in the middle, with the "most exposed and affected" at one end and the "least exposed and affected" at the other. To protect those in the "average" category would require one level of control and cost; but to protect those in the "most susceptible" end would require more.

This issue also came into play in the more esoteric debate over the "value of life," which was fundamental to the economists who sought to reduce public choices to monetary values. Cost-benefit analyses inevitably focused on the value of the lives saved, and this in turn raised the question of whose life was valued at how much. Was the young person at age 5 or 20, for example, worth the same or more than the older person at age 65 or the elderly person at age 80? Was the executive whose income was one million dollars a year worth more in these calculations than the person whose income was $20,000 a year, or was the employed person worth more than the unemployed? Was the person who was genetically susceptible to disease worth less than the person whose immune system was much more resilient? Moreover, how were inequities between generations, the obligation of present to future generations, to be taken into account? Such were the complexities of environmental equity that confronted those who advocated reducing costs by reducing benefits.

Political equity ran through the entire gamut of environmental issues, from inequality to participate in administrative decision making, to inequality of resources for financing media coverage and research, to inequality in the ability to hire staff and finance organizations. In almost every facet of environmental affairs, those who advocated environmental objectives held far fewer political resources than did their opponents. Their influence was especially weakened by the shift in the context of environmental decision making from public debate to administrative action. This shift highlighted a most significant dimension of environmental inequality—the relative economic resources required to influence the outcome of public environmental decisions.

13 ENVIRONMENTAL TECHNOLOGY

Technological innovation was one of the major impulses in environmental engagement. Many existing technologies pollute communities by emitting wastes to air, land, and water. How to develop and adopt more environmentally acceptable technologies? The issues are far-reaching, ranging from extraction and processing of raw products to chemical manufacturing and the agricultural use of pesticides and excess fertilizer. Resource-extraction and land-development technologies progressively disturbed more land more severely with practices such as clearcutting in timber harvest or machine removal of land in strip mining. Increasingly large power plants concentrated waste heat, which had an even greater impact on the water bodies into which the heated water was discharged; larger commercial establishments such as malls concentrated vehicles and hence increased air, water, and land pollution. Ever increasing areas of land paved for parking lots and roads led to downstream flooding and salt pollution of waterways.

Thus, one facet of environmental engagement that was carried out with much heat was the continual face-off between those who defended existing technologies as causing minimal harm and those who argued that new technologies were quite feasible and sought to speed up such innovations and their application. Just as the citizen environmental organizations were at the forefront of environmental science discoveries, so they were at the forefront of the development of new environmental technologies. And just as the affected industries were prone to belittle evolving environmental objectives, they were equally prone to belittle ideas about new environmental technologies that came from others.

However, cracks in that resistance became evident as certain sectors of industry moved forward while others did not. Two major influences led to innovations within the industrial community. One involved competition from those sectors that were adopting new technologies with success and profit, which the conservative sectors found undermined their market share. The other involved government programs to encourage technological innovation such as establishing standards that created markets for new technologies or financing research that could then be applied by innovative industries. Movement toward more acceptable environmental technologies was slow but steady.

Several types of technology were at issue. One of the earliest was technology to treat waste prior to its discharge into the environment so as to reduce its impact. In sewage disposal this came to be known as "wastewater treatment," and the series of stages of treatment came to be called "primary," "secondary," and "tertiary." In air quality, technologies known as "air cleaning" had developed over the years, most of which were intended to remove dust or particles from the air before discharge or to precipitate gases such as sulfur dioxide so that they could be removed as solids. In the case of motors powered by fossil fuels such as gasoline and diesel, the thrust of innovation was a "cleaner burning" engine, and considerable progress was made in this direction.

To the environmental community, the preferable technologies were not those that treated waste before discharge but those that produced less waste in the first place. Improvements in gasoline motors that reduced pollutant discharge were far more preferable than those that treated pollution at the end of the tailpipe. In some cases this meant "alternative" technologies, such as finding less-polluting sources of energy, technologies to make solar energy more feasible, or integrated pest management, which would reduce considerably the need for pesticides in agricultural production. Organic farming was even more preferable because it used technologies that did not require either pesticides or commercial fertilizers. Equally important were those technologies that reduced waste discharge by recapturing and recycling chemicals in the production process, an innovation that came about more readily in some industries because it reutilized a raw material and reduced material costs.

The drive to make technologies more environmentally efficient met with some success as a result of three sets of circumstances: (1) environmental organizations kept an eye on the possibilities for innovation and brought them to the attention of both industry and policymakers; (2) resistance from those within the industrial community using existing technologies often was marked by divisions between those who were more willing to innovate and those who were not; and (3) government agencies developed an extensive fund of engineering expertise and therefore often negotiated directly with industry to advance innovations. The interplay among these three forces focused on technology, but because major elements of public policy were involved in encouraging adoption of environmentally beneficial innovation, it had a profound political dimension as well.

ENVIRONMENTAL STANDARDS

Mandated standards, levels of permitted emissions of air and water pollutants, established the goals of public policy; the sources of pollution, then, were required to apply technologies that could meet the standards. Usually they were given ample lead time to comply. Often the decision to establish the standard bore a close relationship to the technological possibilities, so that the technological innovations were quite feasible. Often they were already in place in some industries that were more technologically advanced, and the task was to generalize them to other sectors of the industry. Hence the role of the standards was both to make existing advanced technologies more widespread and to foster innovation at an earlier stage that might lead to an even greater reduction in pollution levels.

In the Clean Water Act of 1972, secondary treatment of sewage was established as a standard. This meant that treatment had to go beyond removing solids, as was accomplished in primary treatment, to reducing levels of organic material (and its accompanying biological oxygen demand) with secondary treatment. "Biological oxygen demand" (BOD) referred to the oxygen used by biological decomposition of organic material in lakes, rivers, or coastal waters, leaving less available for aquatic life, which therefore would be threatened. In secondary sewage treatment, the BOD of the sewage would be reduced by aeration and oxidation within the treatment plant before discharge. This requirement gave considerable impetus to the technological innovations required to implement secondary treatment. Tertiary treatment aimed at removing dissolved materials such as phosphates, which encouraged algae growth, and toxic chemicals.

And so it was also with air-pollution standards. For years the major concern with air pollution had been smoke, the black particles in the air that reduced visibility, caused household dirt, and were sources of human lung problems. In this area, the main thrust of technological innovation was either to remove the smoke, through what were known as precipitators, or to reduce the potential pollution by pretreating coal, the main source of the pollution, before it was burned. By 1960, therefore, a well-developed industry had arisen to establish "smoke control," and a professional organization, the Air Pollution Control Association, had been formed to bring together administrators, engineers, and industry to develop further technologies. By the 1970s, additional concern for sulfur dioxide emissions had arisen. The smoke-control industry responded by developing technology to chemically combine the SO_2 gas with powdered limestone—"scrubbing" it was called—to produce a solid that would precipitate and reduce the SO_2 emission. After the

Clean Air Act of 1970, emerging scrubber technology became the focal point of several decades of debate over feasible air-pollution technologies.

The debate over the SO_2 scrubbers sharpened several of the larger issues of technological innovation. First was the idea that the best system for dealing with sulfur dioxide was simply to build a tall stack and let the emissions blow away downwind from the community around the coal-burning electric utility, a case of pollution dilution rather than removal. This was the alternative preferred by industry, administrators, and public health experts. However, this argument became outmoded during the 1970s, when it was found that sulfur dioxide did not disappear as it was transported downwind but was chemically transformed into a sulfate that, in the presence of moisture in the air, became sulfuric acid or a particulate sulfate and was deposited downwind as acid rain, snow, fog, or dry particulates that harmed biological life on land and in water. This undermined the preference for tall stacks, fostered a sharper focus on SO_2 removal before emission, and greatly speeded up the development of scrubbers.

The scrubber issue also emphasized the limited engineering skills within the utility industry. Scrubber technology involved chemical reactions and the effect of those reactions on the mechanical equipment that housed them. Hence the crucial problem of reliability, a high level of continued performance, and a low level of interruption in production because of "down time" called for the skills of chemical engineers. But electrical and mechanical rather than chemical engineers were in charge of the utilities, and the knotty problem of achieving a high level of continuous scrubber performance was a new task for them. The first successful application of scrubber technology came when the Louisville Gas and Electric Company placed a chemical engineer at a high level in the company hierarchy, and this individual made the system work. In later years scrubber manufacturers got around this problem by offering to lease scrubbers to utilities and then guaranteeing their effective maintenance by the manufacturing company. Environmental technology often required such shifts in technical skills, which came slowly in more conservative companies.

The scrubber issue also illustrated the direct connection between standards and technological innovation. Installation of scrubbers in the 1970s was confined to new electric power plants; the standards applied only to them. For years extension of the emission standards to older plants was slowed up by the resistance of the utilities. However, in the Clean Air Act of 1990, the reduction of sulfur dioxide emissions in old as well as new plants was mandated by a new approach that established a national cap on yearly sulfur dioxide emissions. A portion of this total, known as a pollution allowance, was then allocated to each plant, which in turn was required to reduce emissions to the level of the allowance. This requirement created a market for pollution-control industries that could help the utilities meet the standards. Producers of low-sulfur coal and scrubber manufacturers competed vigorously for the new market; the scrubber business boomed, and the price of both low-sulfur coal and scrubbers declined sharply.

TECHNOLOGY STANDARDS

Much of the debate over the adequacy of existing technologies revolved around the effects of the pollution: was it harmful or not? That issue often became mired in diverging opinions over the meaning of the factual knowledge, and since most such issues were complex, they were subject to considerable and continuous debate that often resulted in inaction. Such decision paralysis gave rise to the alternative that one should simply apply a technology known to exist or to be on the verge of successful application. This prompted the concept of technology standards, that a desired technology should be worked out for the various polluting industries that, if applied, would reduce pollution to the desired level.

The idea was first applied in the Clean Water Act of 1972, which came after many years of debate over how to achieve acceptable levels of water quality. The act divided standards into two parts. The first group, technology standards, were applied across the board as minimal standards, and the EPA was given the task of setting those technology standards for different industries. Where there were many dischargers into one body of water, it was recognized that the technology standards would not reduce the total discharge to the desired level. These bodies of water were known as "water-quality-limited streams," and for them the task was to identify the desired water quality for that stream and then develop additional controls beyond the technology standards.

To establish technology standards, the act provided a simple strategy: identify the best technologies in place in a given industry, take "the average of the best," and then require that these be mandated as "exemplary standards" for all facilities in that industry. For this model the EPA selected the top 10 percent of performers. The regulated industries, in turn, sued the EPA on the grounds that the wording meant not the top 10 percent but the median of the whole. This, of course, would mean that only 50 percent of firms in the rank order rather than 90 percent would have to apply improved technologies. The courts rejected their claim on the grounds that the EPA was closer to the meaning of the law. Then industry again sued the EPA on the grounds that it had included examples of firms outside the

United States, for example in Canada, to establish its rank order and that the law pertained only to firms within the United States. In reply to this contention, the courts argued that technology was universal in its meaning and that the EPA had acted acceptably.

Technology standards were applied to air quality in the Clean Air Act of 1990 in a manner that intentionally copied the approach of the Clean Water Act of 1972. The issue was the control of toxic air emissions. The act did not leave the "average of the best" to the EPA's discretion but established the top 12.5 percent of firms as the basis for determining the models. This decision came after two decades of debate over toxic emissions. The 1970 Clean Air Act had given the EPA authority to control such emissions, but specific efforts to do so were mired down in intractable controversy over the effects of toxic air pollutants. In order to cut through this decision paralysis, the Congress simply sought to apply technology controls across the board.

In a number of other cases in which it was difficult to be precise about harm and to justify controls on that basis, technology standards were an alternative strategy. One such case involved lead in gasoline. Scientific evidence about the harm of lead that was inhaled from the air, the main source of which was lead in gasoline, made clear that the problem was widespread and of particular harm to the neurological development of children. Once the issue was felt to be important, the immediate proposal was to eliminate lead from gasoline. There was no thought about a "threshold" level of lead in gasoline below which there was no harm, hence the decision was simply to go for a technical solution that would eliminate lead in gasoline.

Another case dealt with erosion from land-disturbing activities. In this instance, the precise damage from sedimentation was not the basis for determining given amounts of controls; instead regulations simply required certain erosion-control activities to be applied, such as establishing erosion barriers around building sites or strip-mining or tree-harvesting activities. Still another case involved the effect of lawn fertilization on lakes, in which strategies to reduce the adverse effect on lakes rested heavily on reduced fertilization or on buffer strips between lawns and lakes. Similar source-reduction strategies were applied to forest practices that produced environmental harm to streams. But these approaches did not require in-stream monitoring to determine if the control measures were effective.

One knotty problem in the application of technology standards came with the attempt to deal with toxic chemicals in drinking water. A wide range of chemicals were present in drinking water, so many that it appeared to be rather unrealistic to tackle and remove them one by one. Instead a relatively new technology called activated carbon filtration seemed to be capable of removing them all and thus came to be the technology of choice in drinking-water systems. But many suppliers balked because of the cost, so the proposal became highly controversial. The EPA sought to soften the transition to a more desirable technology by providing that either activated carbon or its equivalent in results would be required, but drinking-water suppliers rejected the proposed technologies.

Those who sought to reduce pollution found that interminable debates over its impact stymied action, so they preferred to stress direct innovations in production technologies. They emphasized "source reduction" in which new technologies and processes would reduce the creation of pollution in the first place and thus avoid either waste treatment or paralytic debates over environmental effects. This included reducing the amount of initial raw materials used or recycling materials within the industrial processes. It shifted the focus on technologies from those used to remove or treat waste pollution to those capable of reducing its production in the first place through process change and product reformulation so that toxic materials were not used or created in production.

Source reduction came slowly to industry and it seemed to make greater headway in Europe than in America, where it was resisted vigorously by American industry. It was embodied in the first major proposal to control pesticide discharges into streams in which Senator Abraham Ribicoff of Connecticut proposed a bill that would require federal licensing and monitoring of the offending plants. One of the first and best-known source-reduction ventures in the United States came from the efforts of Joseph Ling, vice-president for environmental affairs at the 3M Corporation. He popularized what he called the PPP approach, or "pollution prevention pays," with the argument that companies could save considerable money simply by reducing pollution at the source. He was not well received in the United States and was far more widely known in Europe than in America.

Source reduction shifted the focus on technological innovation even closer to the basic technologies of production and set off a new stage in the resistance to change on the part of industry. They had long avoided a direct focus on their production processes on the ground that to permit policies at that level would interfere with their private rights and expose them to unfair competition from others. They spoke vigorously about "privileged information," which should not be allowed to be made public. Massachusetts and New Jersey attempted to tackle the problem by requiring industries to identify their pollution production and adopt a schedule to reduce emissions rather than simply treating or recycling waste after it was produced.

Source reduction became a central symbolic sound bite in environmental affairs; it was accepted widely as the proper approach, but the regulated industry and the environmental community differed markedly in how to interpret it. To industry, it meant reducing emissions to the environment; hence either waste treatment before emissions or the reuse of material off-site—as, for example, a fuel in energy production—were legitimate measures in source reduction. But the environmental community, often joined by public environmental managers, sought to focus more precisely on reduction at the source, and in a five-tier set of priorities argued that reduction, reuse, and recovery should precede recycling and disposal as the set of priorities. Hence in industrial discharges the key element was a change in production technology and not the enhancement of recycling or waste-reduction after discharge.

The New Jersey source-reducing effort focused sharply on the conservatism of plant staff who resisted innovation in favor of more traditional and familiar ways of operating. It seemed to confirm the key argument in Ling's approach, that success depended heavily on vigorous leadership within the firm and active cooperation from plant staff. Yet what appeared to the environmental community to be simple wisdom in attempts to improve technology met with vigorous resistance from industry, and the Massachusetts and New Jersey programs did not spread to other states. Proposition 65 in California took another tack on source reduction: the requirement that products had to be labeled if they contained chemicals that were known to cause cancer or adverse reproductive effects. Faced with the potential of litigation over failure to label, industries became more open to changes in their process.

The idea of source reduction evolved slowly over the years but was fostered by the fact that economic growth led to more total pollution, even though each source was more environmentally efficient. Automobile emissions were the most dramatic example. While the catalytic converter had cut emissions from cars, on average, by half, miles driven had doubled, resulting in little change in the total production of emissions such as nitrogen dioxides from automobiles. Or though some reductions in sulfur dioxide emissions occurred at power plants, total emissions remained at unacceptable levels. A major change occurred in the Clean Air Act of 1990, in which the focus was not the emissions from each utility source but the total in the nation as a whole that should be reduced. This placed more pressure on improved technological performance, since each technology was now evaluated in the context of an overall national cap. The continued problem of nitrogen oxide emissions from automobiles, trucks, utility plants, and other sources led gradually to a similar approach, and by 1996, one state, Massachusetts, had established a cap for nitrogen oxide discharge for its utilities.

TECHNOLOGY PERFORMANCE: ENVIRONMENTAL MONITORING

Environmental monitoring was fundamental in creating more environmentally efficient technologies. As technologies changed, not only could the detrimental effects of industrial processes be measured but also the emission reductions. Monitoring was time consuming and costly, and although administrative agencies and environmental organizations sought to place the cost of monitoring on the pollution sources, the latter argued that such costs should be borne by the public. At the same time, monitoring emissions such as radiation from nuclear power plants or water acidity or alkalinity in streams gave rise to voluntary citizen monitoring that played an important role in gathering environmental data. Understanding pollution effects also required that ecological change be monitored, such as alterations to fish and invertebrate populations and nesting success or failure. No matter who carried out the monitoring, the task promoted continual improvements in monitoring instruments and biological indicators of harm.

The initial focus of environmental monitoring was to provide information as to compliance: did sources of pollution comply with the regulatory requirements? Did emissions from power plants meet the test of permitted levels; were discharges from sources of wastewater allowable? Early devices for determining these levels were rather crude, such as a simple visual test for smoke in air pollution or a few water-quality chemical tests. As time went on, however, measurements became more varied and extensive. If required simply for general compliance, emissions measurements could be made only at monthly intervals, but technology performance required more frequent data. Hence the Clean Air Act of 1990, which emphasized actual quantities of emissions, mandated continuous emission monitoring, which meant much more frequent testing. Wastewater effluent from industries was subject to more comprehensive chemical measurement. But even this became less preferable than "whole effluent" monitoring, in which the entire effluent with its mixture of chemicals became a preferred measure of effects. It also came to be felt that those effects should be identified in terms of the impact on aquatic organisms, which came to be indicators as important as the chemicals themselves.

Both industry and regulators were often interested primarily in data that would demonstrate the degree of compliance. The benchmark was not environmental change but the terms of the permits as to allowable emissions. The Clean Water Act of 1972 required that both allowable and actual discharges be made public on a regular basis. This provision had resulted from an intense debate in the late 1960s over the secrecy of water-quality information under previous state programs.

Monitoring air quality became more extensive as the issue of toxic air pollution aroused greater interest in the late 1970s. Industries were required to report regularly their toxic discharges from all sources, into both air and water, a program known as the Toxic Release Inventory, or TRI. Communities had a right to know what toxic chemicals they were being exposed to, and this could be done through annual industry reporting. The data was computerized and made available to those who were able to access it and was used often by communities to assess exposure from industry. As time went on, the EPA expanded the categories of industries required to report. Reporting itself was controversial, as the environmental community questioned its accuracy and the industrial sources resisted the requirement.

The TRI gave rise to increasing efforts to extend monitoring to production. The New Jersey source-reduction program involved continual monitoring at crucial points along the production process so as to identify just where improvement took place. This, so the argument went, was no different than traditional monitoring by industry for financial and materials accounting. Now the same scrutiny would be applied to waste management. However, industry objected vigorously to this extension of monitoring on the grounds that it would reveal privileged information or "trade secrets." Yet the pressure on industry continued to focus both on source reduction in the production process and on procedures to measure performance.

Two issues that evolved rapidly in the 1990s centered on this problem. One involved "environmental audits" and the other an international process known as "ISO-14000." Environmental managers had long advocated that industries should audit their internal environmental performance and, especially, that they should reveal the results. Firms arose to conduct audits for industries, and from this two issues emerged. Could an agency use such information to conduct enforcement against firms, and was the information available to the public? Industry sought state laws to legitimize environmental audits but with provisos that would protect audit information from enforcement action and would enable the firm to keep such information from agencies and the public. To the environmental community, this seemed to be just another issue of secrecy in which industries could gain some public relations advantage from their actions but, in effect, remain at liberty to avoid cleanup.

The ISO-14000 program arose from an attempt to develop international industry performance standards. Such standards, which emerged from quite different industry-wide concerns such as materials quality, began to focus on environmental matters. The ostensible objective of these programs was to develop a more "level playing field" among international businesses. But the environmental proposal raised the two conventional issues: just what measures of performance would be agreed on, and how much of the information would be available to regulators and the public? The issue turned on availability of "independently verified performance information." Some leaders in the chemical industry argued that such a policy was the only way industry could obtain the confidence of the public, but others warned of the dangers of such an open and demonstrable process. Thus opposition to independent verification continued.

The Clean Air Act of 1990 promoted marked advances in air-quality programs; among them were steps to tighten loose elements in existing administrative practices and to expand controls to a wider range of sources. The program had identified both the range of the problem and the contributions to it from many specific sources as well as crucial elements for improving performance. Undergirding such a broadened program was expanded monitoring that reflected technical advances in monitoring. However, those who were required to engage in more effective monitoring vigorously resisted.

Monitoring consumer goods gave rise to its own peculiar opposition and debate. Automobile inspection and maintenance was the most well publicized case; mandated safety inspection was well accepted but not emission inspection. Emissions from older vehicles were a major source of air pollution, and a major strategy to reduce emissions was to improve vehicle maintenance. Regular inspections had been required in earlier legislation, but under the Clean Air Act of 1990, the EPA sought to extend these with regard to both the areas covered and the manner of inspection. The agency argued that inspections and repairs should be done by separate firms and that more advanced monitoring devices should be used. This led to intense resistance from the repair shops and the American Automobile Association (AAA). The AAA launched a major campaign of opposition, aroused its automobile-owner members, and succeeded in reversing the program in most states.

Environmental affairs enlisted a considerable amount of citizen interest in voluntary monitoring. Widespread environmental circumstances were difficult for administrative agencies to track, and it was equally obvious that the private business sector was not interested in monitoring at any level beyond their immediate regulatory needs. One form of citizen monitoring, bird counts, had long taken place under the auspices of the Audubon Society. For many years that data had remained rather obscure and untapped and was considered to be too amateurish to be reliable for research or management. However, as the twentieth century wore on, it served increasingly as the basis for scientific studies and management.

A new stage in environmental monitoring developed when water quality engaged the interest of schoolchildren and teachers, who used water quality as a basis for instruction in environmental science. It also involved adult citizen groups that took an active interest in the water quality of their community. Citizens organized around lakes or watersheds in which they were interested and provided considerable voluntary labor to keep tabs on the quality of the water. Often they obtained monitoring equipment and, in the case of the larger lakes, acquired boats that served as classrooms as well as means for frequent monitoring. Participants in these projects around the nation came together, shared their experience, and established a national monitoring organization with its own newsletter. At the same time, monitoring began to be extended to water-related organisms such as salamanders and frogs, and to forests in which the growth of trees and the presence of spring wildflowers became significant subjects of monitoring projects—all serving as indicators of ecosystem health.

INNOVATING AND RETARDING FACTORS IN ENVIRONMENTAL TECHNOLOGY

Public agencies played a central role in environmental technological innovation, while the private sector played a subordinate role. Few innovations came solely and directly out of the marketplace; those that did were the product largely of the growth of consumer demands, such as in organic foods and other "green consumer" wants. Others came as a result of litigation that rendered industries financially liable to those harmed. Most came through the influence of the requirements of the law, such as standards in pollution-control programs and home insulation, or public investments in technical innovation, such as solar energy or more energy-efficient appliances and lighting.

From the very beginnings of the pollution-control programs in both air and water, the Environmental Protection Agency developed a major program of technical assistance. Some of this was indirect, such as attempts to facilitate the dissemination of technical models and knowledge through various publications—technical transfer it was called. Other assistance was more direct, such as when the EPA sought to work with a given plant or a given industry to promote the application of more environmentally acceptable technologies. The Clean Water Act, with its emphasis on technology standards, brought the EPA more forcefully into the realm of setting the direction in technical innovation because it required the agency to assess the state of existing technology to identify the best examples to serve as benchmarks for others to follow.

This activity led both to relatively quiet and effective results in some cases and to highly publicized roadblocks in others. The quiet negotiations were poorly publicized and reported more frequently in the technical rather than the mass media. Thus it was that the EPA worked with manufacturers to reduce the emissions from outboard motors, which were identified as a major source of air pollution, or with truck manufacturers to reduce emissions from truck engines. A minor but successful effort was their negotiations with the manufacturers of faucets to reduce their lead content. And still another success was the negotiations with the manufacturers of wood-burning stoves to establish nationwide standards of manufacture and air-pollution control. In cases such as these, a major influence prompting industry to negotiate was the existence of stricter standards in some of the states, most frequently California, which persuaded industry to be concerned about the prospect of different technical standards in different states and to opt for a single national standard instead. The contributions of EPA expertise to these innovations was rarely recounted in the public media. More frequent were agency efforts to promote innovations that were resisted by private industry, especially iron and steel manufacturers, the utilities, the automobile industry, the paper industry, and the pesticide industry. In such cases, the problem usually was not the availability of the technology but the reluctance of the industry to accept new capital expenditures or, in the case of pest control, new approaches such as integrated pest management, or its commitment to traditional methods of manufacture. Thus the automobile companies long resisted application of the catalytic converter to tailpipe emissions, the utilities long resisted the requirement for sulfur dioxide–reducing scrubbers, the coke industry long resisted the allocation of labor costs to maintain coke oven doors to reduce escaping gasses, the paper industry long resisted the use of either chlorine-reducing or chlorine-free paper-making technology, and the agrochemical industry long resisted the challenge of reducing the dependence of agriculture on pesticides.

In most of these more intractable problems, two sets of influences bolstered the EPA's continued search for more environmentally acceptable technologies. One influence came from the manufacturers of innovative technologies themselves, such as those who produced catalytic converters or those who produced air-cleaning equipment. These manufacturers touted the environmental benefits of their own technologies and played an important role in both the drive to secure firm environmental standards to provide markets for their products and the negotiations in which the EPA sought to advance their application. The other was the environmental organizations, none of which had the resources to foster technical innovation on their own, but some of which had the resources to keep in touch with the technical environmental world and to throw their weight in favor of a more environmentally acceptable approach. The technical staff of the Greenpeace environmental organization, for example, kept an eye on innovations in Europe and sought to encourage their use in the United States.

Within the marketplace of polluting industries there were influences that facilitated as well as retarded technical innovation. When innovations simultaneously reduced costs and advanced environmental objectives, they came rather easily. One of the most widely touted was energy efficiency. Especially noted was the waste of energy in factories. The potential for improvement led to the growth of new energy consulting businesses in the 1970s that drew up energy efficiency plans for many an industry. The results constituted one of the major accomplishments, along with household insulation and weatherization, of the entire drive for reducing waste in energy use.

Raw-material recovery in manufacturing held out similar opportunities for increased materials efficiency, but they were far less dramatic and were often limited to specific requirements of specific industries. One that was highly publicized concerned the use of polyvinyl chloride gas in the polyvinyl plastics industry. The 1977 Clean Air Act had greatly restricted emissions of this gas because of the higher rates of cancer among workers in that industry, and when the EPA drew up the resulting regulations, the industry objected that they would be economically destructive. The industry relented when it found that adopting technology to recycle the gas enabled it to use a product formerly wasted and reduced the cost of its raw material.

Opportunities for technical innovation often had to be stimulated by public policy. Many an innovation went slowly until public funds were granted to the sources of pollution. The most widespread case involved federal grants to municipalities for sewage treatment plants. Few cities advanced their treatment works without federal funds. Stationary sources of air pollution were enticed to install air-cleaning equipment either by outright grants or by help with municipal bonds. More extensive was the way in which federal standards provided markets for innovative equipment in such fields as water and air pollution. Once a standard was in place, then a host of private market sectors were stimulated to act: inventors, engineers, entrepreneurs, finance and marketing specialists.

The response to pressures for technical innovation was shaped also by the different competitive roles of small and large businesses. In some cases, environmental regulations produced a situation in which larger firms could absorb the costs while smaller ones could not. This was the case in the attempt to phase out leaking underground storage tanks at gasoline stations. Over the years, what at one time had been a mom-and-pop business—often as an adjunct to an existing grocery store—came to be a business organized by the oil companies, and these were now able to fund the new tanks or had the expertise to know how to take advantage of public funding programs. The problems with technical innovations encountered by small firms were emphasized particularly by the dry-cleaning industry. Here the desire to phase out traditional methods of pollution-producing dry cleaning led to special attempts by environmental agencies to help the industry develop and apply new and less environmentally damaging techniques.

Innovation often was pressed by the unavoidable facts of limits. To only reduce the concentration of harmful chemicals from one source would not reduce the total level of emissions, as increased production or an increase in the number of sources would easily increase total emissions. To shift to the strategy of limiting total amounts placed greater pressure on technological innovation that could meet the new circumstances. The Clean Air Act of 1990 was based on this recognition of limits; it required that sulfur dioxide emissions be reduced and capped. A finite upper atmosphere made clear that there were limits to the increasing concentration of ozone-depleting chemicals. The issue of lead in gasoline led to the conclusion that the technological alternative was not to control lead emissions but to prevent them in the first place—reduce emissions by reducing their sources.

By the end of the twentieth century, this context for environmental policy was making its way into water-quality programs as well. It was not enough to limit the discharge from individual industrial or municipal sources of pollution, because the stream or lake into which the discharges were made was limited. Hence each waste body had a total acceptable maximum load. In the larger environmental context, the "critical load" identified the point at which human-generated chemical deposition into the environment, whether in water or land, was too much to maintain the desired level of environmental quality.

Technologies might well not be adequate to reduce human environmental loads sufficiently. They could achieve much, but perhaps not enough. Even the most advanced environmental technologies would be offset by more people using more resources and imposing more waste on the environment. Some economists thought in terms of the limits imposed simply by the cost of cleanup, and they argued that the cost would soon be far greater than the economy could sustain. But environmental thinking went beyond the economists' predisposition to think in terms of monetary limits to consider the biogeochemical limits indicated by the scientific data itself.

14 THE STRUCTURE OF ENVIRONMENTAL POLITICS

Environmental politics made its own distinctive contribution to the structure of American political institutions and especially to the regional and party patterns of political impulses, the relationships between the varied branches of government, and the forms and degrees of political participation. None of these involved basic changes in the structure of American governing, but each contributed to incremental changes in long-established governing institutions. There were distinctive regional and party patterns in the degrees of support for and opposition to environmental objectives. There were distinctive ways in which the struggle for environmental influence shaped the relationships among executive, legislative, and judicial branches of government on the one hand, and local, state, and national governing institutions on the other. And there were distinctive ways in which public interest in environmental affairs shaped both ideas about the public good and the practice of civic responsibility.

REGION AND PARTY

Some regions led in advancing environmental goals and others lagged behind. This produced particular regional patterns of environmental ideas and practices and of environmental politics. This variation was in sharp contrast to the relative uniformity throughout the nation arising from the pressures of population and consumption that were imposed on the environment, and to the relative uniformity of the continual demand for traditional objectives of economic growth: more jobs, more development, and more monetary income. Regions competed at federal agencies for benefits that would advance them in traditional forms of economic growth, and hence engaged in traditional forms of compromise and adjustment to each other's demands. In contrast, quite different patterns evolved in environmental affairs in which some regions advanced environmental objectives while others fostered opposition, leading not so much to adjustments as to controversy that, over the period of the 1980s and 1990s, took the form of intense regional and party combat.

Regional patterns of environmental politics could be observed best in the voting patterns in the U.S. Congress. From 1971 onward the League of Conservation Voters compiled voting scores for both the Senate and the House at the federal level, and many similar compilations were carried out for state legislatures as well. Compilations of these annual charts over the years displayed a remarkable consistency and stability. Those legislative districts that scored high in one legislative session did so also at the next and throughout the years; those that scored low did so regularly. Constituencies high in environmental support from their legislators at one time continued to score high even when others represented those constituencies, and likewise for those with lower support. Factors inherent in regional society, economy, culture, and political history were major sources of the differences.

Although voting records in the U.S. Senate provide some indication of regional patterns, voting in the U.S. House of Representatives is far more precise, since it involves 435 geographical units spread across the country rather than just 50 states and thus provides for more finely graded voting variations. Those areas at the higher end of the scale of environmental support include New England, New York, New Jersey, the Upper Great Lakes states, Florida, and the West Coast. Those at the lower end were the Gulf Coast states, the Plains states from North Dakota to Oklahoma, and the Mountain states. States in the Upper South, from Virginia to Missouri and Arkansas, and in the Midwest, from Ohio to Iowa, Minnesota, and Pennsylvania, were in between.

These patterns are consistent with more qualitative observations. In the stronger areas, more public resources were devoted to environmental programs; here there were more vigorous and active citizen environmental organizations, more extensive reporting of environmental affairs in the media, more extensive environmental curricula in colleges and universities, more attention to environmental education in primary and secondary schools, and a more extensive selection of environmental books was available in area bookstores. The weaker areas displayed the opposite tendencies. Within these general patterns local variations occurred. In the weaker regions, the stronger areas of environmental support were in cities, such as Salt Lake City, Utah; Boulder and Denver in Colorado; Kansas City, Kansas; and Houston, Dallas, Ft. Worth, and San Antonio in Texas. And within the stronger regions there were marked areas of environmental weakness, most frequently in rural areas. Cities were not all at the same level of environmental culture. Boulder, Colorado, for example, stands on the stronger side and Pittsburgh, Pennsylvania, on the weaker side. The number and content of environmental books displayed in local bookstores is a useful qualitative indicator of a city's level of environmental culture.

A strong economic context underlay these patterns. Regions of lower environmental support tend to be those in which extractive industries, including agriculture, have long played a significant role in the region's economy. Those that still do have an active anti-environmental constituency and those that did formerly but where extractive industry has declined have a legacy of the past that still influences ideas and attitudes. Regions of higher environmental support tend to have newer economies in which service, high-technology, and consumer-oriented economic activities dominate. The first display a wide range of older, producer-based attitudes and values, and the second display newer, consumer-based values. The producer economies and culture tend to support values that stress physical strength and dominance, and the consumer economies and culture tend to support values of nurture, human development, and quality of life.

The voting scores show a sharp difference between the Democratic and Republican Parties. Democrats in the House voted about 2–1 in favor of environmental measures, while Republicans voted about 2–1 against them. The regional variations within the Republican Party were quite similar to the regional variations within the Democratic Party, though at a lower level. Thus Republicans in the stronger environmental regions had higher environmental scores and those in the weaker ones had weaker scores. The same regional variation was true for the Democrats. This reflects the interactive role of constituency and party, that though there was a strong variation based on constituency, there was also an important variation based on party.

Over the years the patterns remained remarkably stable, but a few changes took place. One was a regional increase in environmental scores in the South Atlantic states from Virginia to Florida. As this region became more urbanized, it also had persistently higher environmental scores. Environmental scores in Florida, for example, rose steadily, as did those of Virginia, and Republicans in Florida had scores higher than areas of the Deep South. In both cases, as well as in the Carolinas and in Georgia, the urbanizing regions had the stronger scores and the more rural regions the weaker ones.

An even more noticeable decline took place in the environmental strength in the Republican Party. This occurred in two phases. In the 1980s, during the Reagan years, a divergence took place between the areas of greater and weaker environmental strength within the party. Those of stronger support continued and even increased their strength slightly, but those of weaker support reduced their strength remarkably. As the party began to increase its representation in the House, its agenda became increasingly anti-environmental. In earlier years, the two sections of the party with very different environmental records had coexisted amicably, but now the anti-environmental wing of the party began to dominate its leadership and called for a firm partywide anti-environmental agenda. It placed in positions of party leadership Republican senators and representatives with very low environmental records, many of them with scores of zero, all of which gave rise to a vigorous anti-environmental drive in the 104th Congress elected in 1994.

The Mountain states displayed some contradictory patterns. The dominant legacy of that region was an extractive economy. Yet in the 1970s and thereafter, it attracted many people who came to experience the high quality of its environment, its mountains, lakes, and streams and the opportunities they provided for outdoor enjoyment. Hence the old and the new coexisted in the Mountain West and competed vigorously for influence. The Republican Party dominated the area's politics and expressed strongly anti-environmental attitudes, yet there were significant pockets of environmental strength. Cities began to shape the region's culture; the agricultural colleges, which had long been bulwarks of support for the extractive economy, provided increasing scientific and cultural support for environmental objectives; and citizen environmental organizations began to exercise influence in state and regional affairs. This internal transformation, which caught the Mountain West between past and present, shaped its paradoxical politics.

GOVERNING INSTITUTIONS

Environmental politics worked its way through the nation's major political institutions and simultaneously was shaped by them and also helped to shape them. There were two sets of such institutions. One, referred to by the phrase "separation of powers," involved legislative, executive, and judicial institutions that carried out different governing functions: making laws, executing laws, and interpreting laws. The other, referred to as federalism, involved relationships between state and federal governments; however, the growth of cities brought a new political entity into this vertical structure of government, as did the similar growth of local governments in less urbanized areas. In terms of their formal constitutional role, both urban and local governments had no independent existence under either federal or state constitutions. But in terms of the real world of governing, they had become significant elements of a four-layered system of federal, state, urban, and local government. At the same time, the federal executive branch evolved into two distinct governing entities, the administrative agencies on the one hand, and the executive office of the president on the other, each with its own rules of the game and its own relationship with the Congress and the courts.

Separation of Powers

Although the three main branches of government were usually described in terms of their broad functions of making, executing, and interpreting laws, over the two centuries of American history each had, in fact, been even more a focal point of political competition. Each served as an opportunity to secure political advantage. When one avenue of action provided by one branch of government did not lead to the desired decision, proponents of a given policy sought to use another branch that might lead to better results. Each branch of government had its own particular processes that shaped political participation and its own form of relevant political power.

The most significant change in these governing institutions in the twentieth century was the growth of administrative agencies. Almost every new statute involved the creation of a new administrative apparatus to implement it. The Congress passed general statutes but left the details to the agencies, and they, in turn developed the required technical expertise to handle these. As a result, Congress found that it was difficult to track the administrative results of its own policies and hence sought to assume "oversight" through formal hearings about the results of administration as well as legislation. As the agencies developed their own strategies to implement laws, the executive office of the president had difficulty in establishing some degree of supervision over the whole; it did so primarily by shaping administrative appointments and the budget. Those dissatisfied with decisions by both Congress and the agencies frequently took their case to the courts, which could well overrule both. But as issues became more technical, the courts often argued that they were not able to judge technical accuracy and therefore deferred to agency expertise.

Both those who advocated environmental objectives and those who opposed them had to wend their way through this maze of governing institutions. For environmental opponents, this was all relatively conventional; they had used the full range of governing institutions for years—in fact had helped to shape their power and authority—and hence merely transferred that expertise to the new field of environmental affairs. Their major weapons were lawyers, technical experts, and public relations specialists. The public relations experts were useful primarily when a legislative issue or a highly publicized administrative decision was at hand; their own technical experts were most useful to shape the thinking of agency experts. The lawyers dealt with policy, either in shaping legislation or administrative decisions or in bringing appeals from agency decisions to the courts.

For citizen environmental groups, participation in the politics of governing required learning the strategies of involvement for the first time. At the initial stages in the 1950s and 1960s, environmental organizations confined themselves primarily to lobbying Congress, and their main strategy was to arouse the public through their membership publications or through the print media. As the issues became more technical, both in terms of the substance of the issues and the complexities of administration, they added some technical experts to their staffs. Several new organizations arose to provide legal expertise, such as the Natural Resources Defense Council, the Environmental Defense Fund, and the Sierra Club Legal Defense Fund (not connected with the Sierra Club). The first two of these brought together both experts in environmental subjects and attorneys to try to cope with these evolving governing circumstances. As citizen state environmental organizations sought to cope with state environmental affairs, a few of them, too, were able to hire specialists. Despite these innovations, however, the resources available to environmental organizations to cope with institutional complexities of policy and politics were far less than those available to the environmental opposition.

The main strength of citizen environmental organizations lay with the voting public, and this, in turn, gave them more clout with the Congress. Hence much of their activity consisted of mobilizing the public to influence the course of legislation. They applied this strategy also to the agencies when it came to general administrative policymaking and when seeking to influence the executive branch. As time went on, both the Congress and the executive office of the president tended to become involved in specific decisions rather than general policy, such as a specific appropriation of money, an action to protect a given natural area, or a decision such as a new air-quality standard. When congressional environmental support declined and turned in favor of the environmental opposition, as happened with the election of a Republican majority in 1994, citizen environmental opportunities declined sharply, but the environmental opposition in the 104th Congress also energized the environmental organizations to use more vigorously their traditional strategy of arousing the public, which partially succeeded in deflecting that opposition.

Environmental affairs modified the role of the courts. The issues they raised were not, in principle, new decisions; they involved the age-old problem of where to draw the line between the use of private property and the adverse effects of that use on the public at large. In the past, issues of that sort were dealt with under the common law and had been brought as individual cases to the courts by the aggrieved party. But in the environmental era, a number of these types of issues had been translated into statute law, which defined particular categories of environmental quality, such as air and water, as desirable public objectives and hence gave rise to a full-fledged administrative system to implement new programs. Now the work of the court involved both the interpretation of the statutes and the workings of the administrative system to implement them.

The courts made two major contributions to the complex of environmental decision making. One was to take the world of environmental affairs seriously. At the same time, courts tended to interpret environmental laws narrowly rather than broadly and to limit their application. This was the case with what many thought of as the centerpiece of the spate of new environmental legislation, the National Environmental Policy Act (NEPA) of 1969. That act anticipated a large governmental influence through its provisions requiring agencies to advance environmental objectives. But the courts confined those provisions to procedural rather than to substantive requirements. Hence the issues surrounding the impact statements emphasized whether or not an agency had conducted a comprehensive and interdisciplinary review of a development proposal. Once the court was satisfied that the review had been thorough, then the agency could make any decision it wished to. This only enhanced the substantive power and authority of the agency.

Equally if not more important was the role of the courts in supervising "fairness" in administrative and judicial matters. They acknowledged citizen environmental organizations as legitimate actors in litigation by giving them "standing," the right to bring lawsuits to defend their environmental interests. Property owners had long been able to enjoy standing on the grounds of their property interests. Now environmental attorneys brought to the courts the contention that they could defend an environmental "interest." Federal courts agreed. While this opened the courts to environmental litigation from the citizen side, the environmental opposition used the courts far more frequently and commanded far more legal resources as well. By the year 2000, in fact, the opposition had been able to persuade the courts to restrict citizen standing considerably.

Federalism

Environmental politics became an active element of public affairs at each level of government. Local environments became the subject of intense dispute in local governments; cities had their own "environmental problems"; the states took up active environmental programs, some of which were instituted to carry out the requirements of federal laws, but many of which came out of their own statewide initiatives; and leadership in many environmental areas came from federal action as well. These interacting realms of environmental governing gave rise to many a controversy among the various entities. At the same time, the governing structure gave many opportunities to both environmental advocates and the environmental opposition to advance their objectives by using one level of action to overcome or to thwart another level. It was impossible to understand or to act intelligently within this complex of environmental governing without taking into account the whole.

Many observers sought to simplify this complex of governing institutions by focusing on traditional ideas of federalism and especially by emphasizing the relative authority of federal and state governments. In the West, land issues were dominated by the large federal landholdings in the Mountain and Pacific Coast states, and from time to time there arose among western political leaders demands that the federal government transfer these lands to the states or sell them to private owners. This issue often sustained quite vehement antifederal ideas and strategies in the West. However, beneath this overlay of antifederal strategies were intense divisions within the West between older extractive and newer environmental objectives, and antifederal sentiment was actually shaped by the struggle for advantage within the West rather than between the West and other regions. Western states owned thirty-five million acres of lands that they administered themselves, and these provoked sharp divisions as to how they should be managed and equally sharp contrasts between state and federal land-management practices. Hence the environmental affairs of the West involved separate spheres of governing, one state and the other federal, that were closely intertwined, with one fueling the other.

In matters of air and water pollution, federal authority evolved initially out of earlier state public health programs that in the 1960s were considered to be deficient. Officials of industrial sources of pollution dominated state air and water commissions and in a joint strategy kept their pollution discharges secret. Hence the Clean Air Act of 1970 and the Clean Water Act of 1972 both established standards far stronger than what had evolved in the states. In addition, the Clean Water Act required that pollution information from industrial and municipal sources be made public in periodic reports. Both laws provided that citizens could sue agencies if they did not carry out mandated legal requirements.

As these programs evolved, opposition to them was lessened somewhat by financial and technical assistance from governmental agencies, ranging from municipalities to states and the Environmental Protection Agency. This was especially the case with the municipal sewage program, in which cities were provided federal funds to apply the new federal standards for sewage treatment. Application of new air-pollution standards was facilitated by both federal and municipal financing programs such as more favorable corporate tax write-offs or municipal bonds. Cities also subsidized industries in their water treatment costs by combining municipal and industrial waste treatment in the same plant.

With the passage of time, Congress sought to transfer more of the cost to the states. At the same time federal contributions to the administrative cost of both state air and water programs were reduced. In the case of sewage treatment, some grants were shifted to loans. Similar problems arose for safe drinking-water programs in which the municipalities had to finance both higher standards of treatment and more extensive monitoring. In response, cities demanded that none of these programs move forward without additional federal financing.

The capability of state environmental agencies varied considerably from state to state, depending both on the degree of the state's environmental culture and its administrative and economic capacity. The stronger states had a higher level of performance than did the weaker ones. In order to prod the weaker states, federal enforcement became stricter. Such enforcement became a major thrust of citizen environmental organizations. In response, many state agencies now sought to be freer from federal requirements, either by being allowed to apply more lenient standards or by being allowed to carry out more lax surveillance and enforcement; moreover, they sought to limit public scrutiny of enforcement actions. In many ways, the increasing state influence in environmental affairs was a reversion to the pattern of the 1960s.

Interstate economic and environmental circumstances tended to enhance federal rather than state authority. One case was the interstate character of pollution. In air pollution, the emphasis on long-distance transport of sulfur and nitrogen compounds from their sources to their deposition many miles away across state lines emphasized the fact that only multistate action could tackle the problem effectively. In some cases, this led to regional cooperation among states, such as in New England, but the cause-effect relationships were sufficiently widespread that federal action became the norm for dealing with both sulfate and nitrogen deposition. In other cases, nationwide markets shaped nationwide policies. This was especially the case involving California, which had long been permitted to develop some of its own standards and did so at levels stricter than the national; industries throughout the country with national markets then sought national standards, though often at less strict levels.

A major strategy used by the environmental opposition was to obtain laws in which one level of jurisdiction preempted the freedom of another level to establish stricter environmental programs. The most successful such federal preemption involved noise; federal law prohibited localities from limiting noise from such sources as railroads, trucks, and airplanes. The agrochemical industry also sought federal preemption over state law and local ordinances to establish pesticide standards and precautionary label warnings. During the debate over the Clean Air Act of 1990, a host of industries sought federal preemption of state actions on air-quality matters, most of which failed, but a few that succeeded were for very specific products, such as industrial paints and coatings, which were often so obscure as to be difficult to track.

The most significant cases of environmental preemption, however, involved preemption of local authority by states. Many environmental issues arose when citizens sought to influence environmental conditions in their local communities. The environmental opposition, in turn, advocated state policies to preempt local authority by establishing state standards and then prohibiting counties and municipalities from creating stricter ones in policies pertaining to such matters as wetlands, air quality, strip mining, oil and gas drilling, forest practices, and facilities site permitting. In the 1920s states had granted localities authority to control a variety of developments through zoning, and now in the environmental era the environmental opposition sought to require states to "take back" this power on a case-by-case basis.

CIVIC PARTICIPATION AMID POLITICAL INEQUALITY

If there was any one effort that ran through organized environmental affairs, it was the drive to persuade more people to become a part of the citizen environmental "movement." People were enticed, cajoled, educated, and encouraged to become active in learning, voting, and supporting legislation and administrative action, as well as to write, call, fax, or e-mail decision makers at every stage of the decision-making process. Added up, all this was a major contribution to a fundamental aspect of the American political system—public participation. It took place amid arguments that public cynicism engulfed the nations' public affairs, that voting participation was declining, that the public did not trust its political leaders, and that they were convinced that they counted for little while those with money counted for much in public decision making.

Several major factors underlay this limited interest in public affairs. One was the degree to which people devoted more of their time, energy, and funds to private affairs, which competed favorably with time devoted to public affairs. Another was the degree to which the intense technical feature of most public affairs meant that few could devote the time and energy to understand them, let alone do anything about them. And still another was the heavy influence of those with money in determining who was elected and appointed to decision-making positions and in shaping the resulting decisions. In addition to these elements of the political system that made for reduced citizen participation, the nation's educational system did little to promote civic participation. In earlier years "civics" had been a major feature of the secondary school curriculum; now it had all but disappeared. And the colleges and universities were dominated by academics who confined themselves to analyzing the political system, if they studied it at all, rather than to encouraging participation.

Amid all these tendencies working to reduce civic participation, the organized environmental movement sought to enhance it. They urged members of the public to support their organizations with memberships and funds; they encouraged them to become active beyond financial support by seeking to influence their legislators and administrators. Above all, citizens were pressed to become knowledgeable about issues and the way government worked so as to know when and how to participate. On the whole, the environmental movement—as did many such voluntary organizations—played an important role in civic enterprise, shaping new kinds of political participation and enabling people to feel that they could "make a difference."

Environmental civic participation, however, took place within the fundamental experience of political inequality. Equality of suffrage had long been the hallmark of political equality in America, and over the years suffrage had been extended to more and more people. But political power had long since gone beyond the limited act of voting and had come to involve a much wider range of considerations such as the ability to command information, to influence the mass media, to finance legislative campaigns, and to command professional skills and legal resources. Environmental organizations were hampered by limited resources and thus demonstrated fully the role of political inequality in contemporary American politics. Although environmental issues commanded broad popular support and were considered to be an integral part of the "public interest," they commanded relatively weak political and financial resources and continually struggled against a more powerful opposition.

Environmental issues were broadly conceived public issues; they represented the interests of a broad-based segment of the public, and they involved environmental conditions that were widely shared. Their relevance as public policy came from the fact that the value of the environment to individuals lay in public policies that would sustain and enhance those values when private action alone could not. Hence they readily took on the description of being "in the public interest." At the same time the anti-environmental opposition readily took on the description of being a "private interest," reflecting primarily the financial interests of private investors. The most vigorous participants in the environmental opposition were private industries and developers, including immigrant agents or employers or forestry and grazing industries, and this played a dominant role in defining the characteristics of the environmental opposition. In the debate over the new clean-air standards in 1997, for example, some 65 percent of respondents in one survey said that if an elected representative voted against the standards, it meant that they were simply doing the bidding of private industry.

The broad support for environmental objectives within the general public meant that, for the environmental community, the main political support on which they could rely was the general public and the voters. But they faced massive resources from the opposition. Some of these took the form of financial support for political candidates, a field in which the environmental community increased its activities over the years but in which it was continually outmatched. Or it took the form of massively supported public media campaigns, which the environmental opposition took up as a major strategy for countering public environmental support. The message was heavily "green," stating either that a past anti-environmentalism had now turned into favorable environmental support and hence the opposition could now be trusted, or that the opposition had a more effective environmental program than did the environmental organizations themselves. The most influential role of opposition media campaigns, however, was in referendum campaigns in which the opposition to state and local pro-environmental initiatives instigated heavily financed efforts to discredit the proposals, and these efforts often succeeded.

Participation in administrative proceedings such as rule making also involved relative political power in the form of commanding professional skills. Environmental organizations could draw on lawyers and technical experts far less than could the opposition. Thus professionals paid by the environmental opposition frequently confronted citizen volunteers in these situations. Legal resources were more than unequal. Court decisions that were favorable to environmental objectives received considerable publicity because of their media drama. Yet it was the environmental opposition that commanded most of the legal talent and was able to bring a host of legal resources to bear on a wide range of issues. Finally, at least a dozen "think tanks" well financed by corporate funds were in the ideological forefront of the environmental opposition (Cato Institute, Free Enterprise Foundation), but there were only a handful of environmental think tanks (Worldwatch Institute, World Resources Institute) that were devoted to similar research and publication.

Environmental affairs were only weakly supported among the nation's institutional leaders, the professions, corporate business, government, the media, and institutions of higher education. Granted, all these elements had come to believe that environmental objectives were in some general fashion legitimate public objectives and they were somewhat willing to affirm the wisdom of past accomplishments, but they were reluctant to move forward toward more advanced environmental goals. They accepted without serious question the nation's developmental objectives in terms of more jobs, greater material productivity, higher consumption, and higher levels of conventional gross national product. One indication of this inequality in the nation's objectives was the degree to which few environmental leaders appeared on the political scene; those who seemed friendly to some environmental objectives were placed under a considerable cloud and seemed compelled to speak of those objectives only as an afterthought amid the more compelling attractions of development.

15 THE RESULTS OF POLICY

Four decades of environmental affairs have led to a wide range of environmental programs and activities that provide an opportunity for assessments of the results. What has been accomplished? Which policies have been more successful and which ones less so? In his book *A Moment on Earth,* Gregg Easterbrook argued that the environmental movement exercised overwhelming influence and sway in the 1970s and 1980s, an exaggeration that seems to be based more on ideology than on careful analysis. For others, none of the varied environmental programs worked; they were a massive waste of effort and funds. Such writers often purport to be the "real environmentalists" who are more "realistic." Or there is the argument that though some programs worked in the early days, they have not worked in more recent times and hence there is a need for a drastic change.

Judgments of administrators, who record their accomplishments with pride but have difficulty carving out directions for the future, are a bit closer to the mark, but they must also be somewhat suspect since agencies tend to exaggerate their achievements in order to enhance their favorable rating. The environmental community tends to have the most prudent view of the results by balancing statements as to accomplishments and failures. One should especially emphasize that even in the heady days of the seventies and eighties there was as much environmental failure as environmental gain, as much successful environmental opposition as successful environmental initiative.

An adequate evaluation should be based on a systematic analysis of results and on the wide variations in the outcomes of many different policies and programs. The approach used here attempts to provide such a framework. It focuses on several hundred such issues, each a distinctive type of activity either to protect or restore favorable environments, each of which can be assessed by itself. Moreover, it stems from analysis of specific rather than more general consequences. Instead of evaluating programs to preserve nature in general, for example, the focus is on more specific facets of nature preservation such as wilderness, natural areas, wild and scenic rivers, endangered species, or biodiversity. Air-quality programs are specified in terms of antidegradation, reduction of air pollution in heavily impacted areas, or exposures to lead and toxic air pollutants as well as conventional ones. Environmental technology achievements are differentiated in terms of waste-reduction technologies in air quality, sewage wastewater treatment, solar energy, or energy efficiency in automobiles.

An initial step in the evaluation is to determine the degree to which an underlying environmental idea has become widely or commonly accepted, if nature preservation is desirable, or if air and water pollution are considered to be problems that need to be dealt with. These goals, in general, are widely accepted. But notions that a growing population or high levels of consumption within the United States are problems are far less accepted. Hence while the first group tends to provide a basis for policy action, the second does not. One element of this limited action comes from the fact that few environmental organizations tackle consumption and population problems because there is only limited public support on which to draw. For more widely accepted issues, more progress is achieved.

There are two very different standards for judging effectiveness. One pertains to legally acceptable performance, whether or not the results conform to the law, a standard used heavily by the agencies and the regulated industry. Environmental organizations, on the other hand, are far more inclined to use environmental circumstances as a standard of judgment: what is the condition of the environment with which a program is intended to deal, and does policy achieve an acceptable environment? Whereas the regulated industry and agencies might well focus on "compliance" as the test of achievement, the environmental community will focus on environmental results.

A significant element in environmental accomplishment is the degree to which an environmental objective has enlisted voluntary as well as public action. Some of the most successful environmental programs are those that rely heavily on voluntary initiatives but have some governmental assistance. These include, for example, the private land conservancies, which marshal private funds and energy but are aided in raising those funds by the income tax deductions from charitable gifts, or glass and newspaper recycling activities, which initially were voluntary and over the years became more public and mandatory. Voluntary water-quality monitoring programs also evolved with considerable private initiative. Many programs depended on governmental action, either as direct administrative activities, such as public land management, or as legal standards of performance, which produce varied reactions from the regulated industry ranging from compliance to indifference and evasion of the law.

Finally, a major element in relative environmental success is the degree to which the environmental opposition played a significant role in retarding achievement. In many cases, such as technological innovation, those managing existing production technologies resisted change and slowed down achievement, often because of decisions to continue profitable production from older and more polluting technologies and to avoid the investments in newer and less polluting ones. Land developers retarded objectives such as wetlands and endangered species protection, the forest industry retarded achievement of environmental forestry objectives, and agriculture retarded efforts to reduce pesticide use or nonpoint source runoff. A major roadblock to environmental knowledge has been the resistance on the part of almost all segments of the private economy to finance environmental monitoring, which provides the basis to judge environmental conditions.

THE GREATEST SUCCESSES

The largest single group of the more successful ventures involves the protection of nature. These range from legally established wilderness areas to land conservation in more settled areas to provision of outdoor recreation in a wildlands setting. Although these programs have met with considerable resistance over the years, they have also enjoyed considerable support, and on the whole they have been one of the most successful types of environmental programs. Several general ideas of wide public acceptance lie behind them. One is that nature preservation is an important element of the desired standard of living in an urbanized society. A wide range of evidence testifies to the high level of interest in nature preservation—evidence ranging from outdoor activities to the purchase of books and equipment to aid in the enjoyment of nature to the popularity of television nature programs.

Nature protection and enjoyment have evolved steadily over the years, from the initial interest in outdoor recreation that originated in the 1920s, grew in the 1930s, and spurted ahead rapidly after World War II. It was reflected in the National Outdoor Recreation Resources Review Commission in the late 1950s, to new programs such as the Wilderness Act of 1964 and the National Trails and the Wild and Scenic Rivers Acts of 1968. The range of interest expanded into many nooks and crannies of both federal and state programs, including wild and scenic river protection at the state level for intrastate streams, identification and protection of natural areas, and creation of hiking trails as many states and regions sought to develop their own segments of the national trail system. In the 1970s, a range of new interests in wild resources emerged to extend interest beyond game animals to encompass a host of nongame animals and plants. The focus was on federal and state endangered species programs, but it went beyond these to "species of special concern"; each state was symbolically associated with a given species, such as the jack pine warbler in Michigan or the common loon in the lake country of the Northeast.

Such concern for so many species came to be talked about in terms of "conserving biodiversity."

Land conservation came to be basic to nature protection, as only with relevant land management would it be possible to protect nature. Hence land acquisition and protection from private-market land forces received wide public support. The general assumption behind such measures was that private landowners tended to be interested primarily in the short-term financial value of their land for development and sale, and hence the only way of avoiding that was to protect land through public means rather than the private market. In many states, bond issues and special taxing arrangements evolved to provide funds for land acquisition, and in most states there was a steady advance in the amount of land added to state parks and protected areas. Even more dramatic were the land trusts that acquired land through gift or purchase and then held the land in trust to be protected from development. These took off in the late 1970s, grew steadily during the 1980s, and by the year 2000 had reached over 1,100. The land conservancy movement elicited considerable voluntary action in terms of funds, time, and energy, and individual conservancies also served as major centers of privately organized environmental education.

Closely related to these were programs organized around a particular species, which combined voluntary support and organization, research, and protection policies. These were a distinctive form of organization that, though limited in scope, were highly successful. Programs to protect and recover bears, wolves, loons, bald eagles, peregrine falcons, raptors in general, bats, butterflies, invertebrates, and a wide range of other specific flora and fauna attracted both popular and scientific support and then played a role in advancing the program for that particular species in private conservation organizations as well as public agencies. These activities tended to link with the economy of wildlife observation, study, and enjoyment. The most dramatic of these were the various centers of research, education, and action that arose for wolves in Minnesota, raptors in Idaho, and bighorn sheep in Wyoming. Once again some of the most successful environmental enterprises seem to involve niches in the world of voluntary interest and action in which people could join with each other to carve out a place for their interest.

The most successful environmental pollution program involved the removal of lead from gasoline. Underlying this drive was the successful demonstration from science that lead had a detrimental effect on the neurological development of children and that it was inhaled not only by children but also by their mothers, who transmitted it to the fetus via the placenta. The science involved an interrelated set of data about exposures, transmissions, and effects that was able to overcome the opposition of the lead industry, its strategy to undercut each facet of the evolving science, its firm disagreement with conclusions about effects, and its opposition to removal. Environmental protection and health agencies sought to extend protection to other lead exposures, such as in drinking water, food, and house paint, which children ingested, and in soil surrounding houses, but results came more slowly because of resistance from those who might be held responsible for cleanup.

The advance of environmental science was one of the major achievements in environmental affairs. Beginning with the early years after World War II, knowledge about the way the environment works accumulated bit by bit. Many disciplines examined many facets of the environment, ranging from changes in ecological systems to biogeochemical processes in transporting a pollutant from its origin to its transmission and effects. In many cases, only the outlines of environmental circumstances were discernible and often new knowledge identified new problems along with the known. Yet, in ways often more fundamental than was immediately obvious, environmental knowledge accumulated to establish more fully the basis for environmental thought and action. By the year 2000, knowledge about air pollution, its sources, its transmission, and its effects, for example, was far more extensive than it was in 1950, and each increment of new knowledge added to understanding it.

These advances did not, however, always have smooth sailing; to understand relative success requires that one understand how influences from the public, the professions, and government agencies overcame the opposition from industry. The lead industry fought hard and long against the elimination of lead from gasoline. From the early twentieth century on, the forest industry opposed parks and wilderness designations, wild and scenic river selections, and a host of applications of nature protection objectives. On the whole, one could argue that a restraining force that continually tended to soften support for these most successful environmental objectives was the general weakness in environmental advocacy from the nation's institutional leaders in government, business, the professions, and private institutions. That relative weakness and indifference was more influential in other environmental issues but was vigorously present in the skepticism about the worth of natural values to society. But this skepticism was overcome by a combination of the support for natural values in the public as a whole and the role of those values as consumer preferences in the nation's economy.

UNFULFILLED OPPORTUNITIES AMID ENVIRONMENTAL GAINS

Many environmental objectives established at an earlier time continued to be implemented amid the new directions of the environmental era. Most of these were resource-based programs dealing with land and water managed in various stages of preservation and protection, which were under continual developmental pressure. At the same time, a few programs that obtained wide support, such as recycling, were marked by partial achievement and moved ahead in a series of fits and starts. Thus, we describe this level of relative success as one of both environmental gains and environmental opportunities yet to be realized.

Some of the nation's most important natural environmental assets were its public lands, long established in the form of national and state parks and national wildlife refuges, to which were now added new protected resources such as wetlands, marine sanctuaries, and lakes, bays, and estuaries. Many of these public resources had statutory environmental rather than developmental mandates, such as the national parks and wildlife refuges and the state parks. Yet developmental pressures compromised those objectives; in the case of state and national parks, it was pressure from those who sought to develop tourist facilities within them, and in the case of wildlife refuges, from those who sought to promote agricultural or extractive industries within them.

After World War II, those with environmental interests in the public lands sought to insulate them more effectively from developmental pressures, but they were only partially successful. State parks, which were established primarily for outdoor recreation and nature protection, came to be thought of as sources of rural economic development, and some states established tourist lodges in them, which evolved into developed recreation areas with golf courses and even convention centers. Visitors to the national parks increased steadily, and the tourist accommodations grew in response, evolving from family-owned enterprises that had long provided park facilities to ownership by international hotel chains, who looked upon the parks as a source of ever increasing business in which profits measured by the larger world of hotel competition were the bottom line.

The national wildlife refuges were one of the nation's most distinctive environmental assets. Managed especially for waterfowl, the refuges served as habitat for a wide range of plants and animals, and in the environmental era they attracted personnel with a broad range of nature protection objectives. But they also were subject to continued developmental pressures, so those who sought to protect their natural qualities tried to obtain a statute—often known as an "organic act"—that would more firmly embed those objectives in law. The two sides remained in a stalemate. In the midst of the gridlock, President Bill Clinton issued executive orders that established the desired environmental direction, and in 1997 Congress approved similar objectives in an Organic Act for the Fish and Wildlife Refuges.

In the years after 1970, several new forms of land and water preservation came into being that established promising new directions but also encountered roadblocks and thus moved ahead only slowly. These new programs included wetlands, coastal zones, marine sanctuaries, and estuaries. Wetlands were one of the most instructive, since they attracted attention as environmental assets only in the last third of the twentieth century, when the values of wetlands were brought into the nation's consciousness. Strategies to protect wetlands on private lands moved forward slowly and with considerable tension. Much enthusiasm evolved about the possibility of restoring wetlands, but this also proceeded slowly and with mixed results. The most significant accomplishment on this score, however, was the development of a comprehensive and firm wetlands science that undergirded action and served as a vantage point from which to defend the program.

Offshore waters as well as the coastal areas became another subject of attention. The first of consequence were the national seashores and lakeshores that became a part of the national park system primarily in the 1960s, each one subject to a new congressional enactment. Marine sanctuaries were intended to protect both the aesthetic and biological features of marine areas; they, too, ran into pressures—from coastal land developers, the fishing industry, and coral-reef sightseers who objected to protection measures. The coastal zone as a whole was subject to one of the few federal land management laws in which the states, under federal guidelines, were to manage an area for both preservation and development. In this program, environmental objectives got a foot in the door but continually struggled against the pressures for development in regions of the nation's most rapid population growth.

Waste recycling entered into the nation's environmental consciousness early in the environmental era; it was especially popular as a consumer venture, becoming a part of many a household ethic, fostered by young people as well as adults as a family enterprise. The success of recycling, however, depended far more on the ability of the business world to make recycling profitable, and this led to continual ups and downs in a trajectory of relative success. Effective laws greatly facilitated the process, as did effective education at the household level. A few cities approached a 50 percent recycling rate, but many others lagged much further behind. Recycling inched forward as niche markets developed for recycled materials and promoters of recycling worked with industries to advance it. Despite the glitches, popular support persisted and one could observe slow and steady increases in the rate of recycling.

The national estuary program seemed to have potential, but evolved slowly. Designations of national estuaries focused especially on their water quality and served to concentrate funds and the energies of citizens and administrative agencies to enhance estuaries as attractive environments for residence and tourism, as well as fish production. The program drew upon a previous focus on bays and lakes, of which the Chesapeake Bay and the San Francisco Bay were prime examples. As the programs evolved and more knowledge about the estuaries and bays accumulated, however, the scope of the problem widened, for example, to include toxic deposition from the air as well as conventional watershed pollutants such as fertilizers and pesticides. These were popular programs that made headway with limited opposition save from adjacent landowners who were called on to curb their land uses that were sources of pollution. In some cases, a sense of larger consciousness even roused cooperation from such owners.

Public involvement helped to keep some environmental ventures on track with slow progress if not spectacular success. Many units of the state and federal park and wildlife systems developed associated citizen groups that provided significant labor to supervise campgrounds, maintain facilities, and raise funds for crucial projects; such groups also often provided low-key political support for their protection. Around wetlands, estuaries, bays, and lakes citizen groups organized with a focus on the particular water resource to foster water-based environmental education for schoolchildren, lead adult educational field trips, undertake monitoring, raise funds, and provide crucial community support. Citizen groups also helped public authorities to extend recycling, especially with special projects such as telephone books, toxic materials, and other items not provided for in the more extensive public recycling ventures. Public involvement varied, but it certainly helped to keep many an environmental venture moving toward some degree of success.

BETWIXT SUCCESS AND FAILURE

A wide range of environmental policies and organizations displayed success in instituting programs but limitations in achieving results. Progress was being made, but there were barriers yet to be overcome. Incrementalism reigned. Each step forward faced a multitude of hurdles as environmental opponents used science, economics, technology, law, and policy to restrain environmental progress. Persistent drives on the part of the public and environmental organizations faced extensive resistance from the environmental opposition.

Most efforts to reduce and control air, water, and solid-waste pollution faced this resistance. In each case, significant laws were on the books, most of which established clear and effective environmental goals, but the implementation was resisted by those regulated industries that were required to change their practices.

The strongest statement of objectives occurred in water quality, where the goal was to make all streams both fishable and swimmable; the law provided that all discharges short of this were illegal but that permits would be issued temporarily until the violations were corrected. Similar goals were not part of the clean-air program. For solid waste, a system of ranking strategies, from reduction to reuse to recycling and landfills, has established some degree of standards.

With all of these initiatives, however, whether goals were fully established or not, the resulting programs became entangled in a mass of pressures and counter-pressures that slowed progress considerably. Water quality in streams and lakes improved, but by the year 2000 there was still a long way to go to meet the fishable and swimmable standards of the 1972 act. High-quality waters were protected more by neglect than by programs, and the very wisdom of the antidegradation program remained the subject of much opposition in the face of which the objective was kept alive only by litigation from citizen groups. The technology standards of the act were clearly insufficient to clean up streams and higher levels of control were required, but the strategy, known as establishing total maximum daily loads for streams, also remained relatively unused and agencies moved slowly, often only under court orders, to implement it. Citizen interest and observation remains a crucial driving force in water-quality improvement in the face of recalcitrance from sources of pollution, and where citizen input is strong, such as on lakes and selected rivers, it plays a significant role in improvement.

Laws to control air pollution became progressively more comprehensive with each new act passed from 1963 to 1990. The 1990 act tightened the regulatory system to cover more sources of pollution and to make implementation more effective. Especially significant was the new approach to sulfur dioxide emissions that shifted from standards in the ambient air to total amounts of emissions, required those emissions to be cut in half, and then established a cap beyond which emissions would not rise in the future no matter what the total number of sources. Although progress occurred in reducing some pollutants, such as carbon monoxide and sulfur dioxide, reductions by half in nitrogen oxide emissions from individual cars were balanced by increased miles driven, so that little overall gain was made in total emissions. By the late 1990s, it appeared that the next step was to place a cap on nitrogen oxides emissions, but the proposal was stymied by the demands of the main two sources, utilities and automobile manufacturers, that the burden of control be borne by the other.

Cleanup of old waste sites was high in the consciousness of the American people, a consciousness that continued to drive the program even though many environmental managers doubted its importance. Progress on the issue hinged primarily on who should bear the cost and this had two twists to it. One was whether the costs should be borne by the industries responsible for waste sites or by the taxpayer. But there was an even more knotty issue: if the industries were responsible, should they or their insurers pay? Cleanup came slowly and often was limited, but the main drama, which frequently occurred in the quiet realm of the courtroom away from the media, was between the insurance companies and the "responsible parties." Their respective attorneys seemed to be the focal point of continued political maneuvering and garnered large amounts of the available funds. The problem received far more publicity than progress.

The rare and endangered species program involved both a high level of public interest and consciousness on the one hand and intense opposition from land developers on the other, so that the protection and restoration program went forward only slowly. The program was implemented in fits and starts, making the most headway when site-specific compromises could be worked out. Landowners, both public and private, were not yet convinced that their ownership included responsibility to preserve and enhance species, and they insisted both that they be compensated for their protection activities and that protection for some species would absolve them of future protection for more. They were especially contemptuous of the biological science that served as the basis of the program and sought to undermine the credibility of the scientists who sustained it.

In the early 1970s three interrelated protection programs—for floodplains, steep slopes, and coastal areas—became quite salient; they all involved human occupation of areas subject to natural hazards and led to the overall concept that any occupation should be "designed with nature." Efforts to restrict occupancy had only limited success; some came up against the peculiarities of law in each state with respect to restrictions on the use of property; some were simply ignored; some were left unattended because of problems of supervision and administration. The driving force that moved the programs ahead was the cost of rebuilding after disasters, often with help from taxpayers through federal disaster relief. As the levees broke in the Mississippi and Missouri River floods of 1993, it became clear that their replacement could not be financed; hence some floodlands were purchased by the federal government and transformed into a wetland-wildlife area and some flooded towns were moved to higher ground with federal funds.

Improvements in environmental technologies also tended to move forward, though with a slow pace and with continued resistance from industries who sought to protect their investment in existing technologies against change. The pace was irregular. Installation of sulfur dioxide scrubbers for coal-fired electric plants, for example, had a rocky and intensely controversial start in the early 1970s. When they were mandated for new power plants and then for all power plants in the 1990 Clean Air Act, cleanup moved ahead rapidly. For automobile exhausts, the catalytic converter brought rapid gains, after a period of opposition from the automobile manufacturers, and the pace then slowed down as further progress in developing more fuel-efficient automobiles went through another rocky period of recurring intense opposition from the automobile manufacturers. As utility emissions became a major focus of air-pollution policies in the late 1990s, attention once again shifted to installation of new pollution-control technologies, this time to selective catalytic reduction (SCR) to reduce nitrogen oxide emissions. The pace of technological development seemed to depend on the advance of federally mandated standards that fostered markets for new technologies, the degree to which private industry and the Environmental Protection Agency worked together to foster new technologies cooperatively, and the impact of innovations abroad, such as in the pulp and paper industry, which tended to force competitive pressures for change on U.S. industries.

FRUSTRATED ENVIRONMENTAL PROGRAMS

In a considerable number of cases, environmental initiatives were established and important initial steps taken, but they were heavily stymied by vigorous opposition or limited support. They usually involved a limited degree of public consciousness, which if heightened might have driven them further; this along with strong opposition led to varied instances in which environmental ambitions strongly outran policy and accomplishment.

One of the earliest of these involved pesticides, which had played a significant role in the evolving environmental issues in the 1960s partly because of Rachel Carson's book *Silent Spring*, partly because of the publicity over a dramatic administrative proceeding involving DDT in Wisconsin, and partly because of sensational episodes, such as a fishkill in the lower Mississippi River due to the discharge of eldrin and dieldrin from a pesticide-manufacturing firm in Memphis, Tenn. But though there was a heightened general consciousness about the potential harm from pesticides, and though the manufacture—not the sale—of a few was prohibited, the pesticide regulatory program was one of the major failures of the environmental era. Judgment as to the harm of pesticides was in the hands of a body in the EPA, the Science Advisory Panel, that was dominated by those friendly to agriculture and to the pesticide manufacturers and rarely included any members specialized in the effects of pesticides on humans or wildlife. Few of the basic biological pathways leading to harm were brought to bear on analysis of effects, including such matters as bioaccumulation, multiple exposures, multiple effects, and biochemical transformation along the routes of pesticide transmission in land and water. These ideas came to regulation only slowly and in piecemeal rather than comprehensive fashion.

A similar fate befell the persistent interest in indoor air pollution. This comprehensive problem came to attention in the late 1960s largely through the initiative of individuals in the EPA. But the enormous range of its implications led to intense opposition from the building, real estate, home improvement, home furnishings, and appliance industries, and the program was squelched by the Reagan administration. It reemerged only as one issue after another—such as radon, lead in old paint, asbestos, and chemical exposures from synthetic materials in carpets, some types of insulation, rugs, household cleaners, and smoking—came to the nation's attention through the media. A patchwork of separate programs resulted, despite the ready availability of more comprehensive approaches, but Congress rarely went beyond a research program to action, and even research was strongly resisted by the industry. Issue after issue was caught in debate over the meaning of the science as one study was pitted against another.

A federal nongame wildlife program was authorized in 1980 by Congress, which contemplated a state matching-fund program, but monies were never appropriated. In the 1990s a drive arose to establish a federal tax on many forms of outdoor equipment used by appreciative nongame users, much like the long-existing taxes on hunting and fishing equipment, which would be distributed to the states in the same manner as those earlier funds. But by the end of the twentieth century, the success of this proposal remained to be seen. Support for fish and wildlife, popular in earlier years, seemed to lose strength as the century wore on and was able to maintain momentum only because of past initiatives, such as the earlier taxes on sport equipment. One earlier program that continued to make significant contributions toward wildlife objectives was the duck stamp, which provided a continuing fund for purchasing wildlife refuge lands for waterfowl. Although programs to protect biodiversity in the Tropics were popular, similar approaches applied to the United States elicited far less support.

The fate of two research areas related to wildlife made clear that the nation's wild resources had only tenuous support if they depended on innovative public programs. One of these was ecological science. In the 1970s Congress agreed to enhance support for ecological research, and the Environmental Protection Agency, which was becoming more interested in the impact of pollution on wild resources, brought the subject more fully into its scientific and regulatory program. Especially important was the establishment of over a dozen centers for long-term ecological research. But the program's weakness was revealed in the degree to which Congress found it vulnerable to funding cuts, thus greatly weakening the possibility of securing knowledge about long-term ecological conditions.

In a closely related subject, the application of ecological science and concepts to forest management, progress was even more limited. Professional foresters and forest-management agencies were heavily committed to wood production and resisted the application of such ecological objectives as biodiversity, the creation of older forests, and ecosystem description, analysis, and management. As the environmental community and ecological scientists brought these approaches to forest professionals and forest mangers, they met steady resistance.

The elimination of billboards from highways, long an environmental objective to advance highway aesthetics, went through a rocky history. In the 1960s efforts to require interstate highways to be free from billboards had some success, but over the years this program was eroded, and because of the political strength of the billboard industry, it was difficult to extend it. There were vigorous encounters such as the efforts by billboard companies to cut down trees that obscured the billboards from motorists. One environmental organization, Scenic America, took up the issue and helped many cities to reduce or eliminate billboards, but progress was slow and depended on the circumstances of particular cities. Arguments as to jobs and economic growth frequently outweighed those that emphasized improving the environmental quality of the city.

Environmental affairs brought to the field of economics new ideas and new approaches that worked environmental activity into the economic equation, but again the progress was quite slow. Environmental consumption, such as the market for goods and services in outdoor enjoyment, the higher value of real estate in more natural and forested areas with strong aesthetic components, or the purchase of more efficient energy equipment or equipment to reduce pollution, could be thought of as major elements of the economy. These factors could contribute to either production or consumption in an environmental economy, a significant sector of the larger economy. Some environmental economists argued that one erroneous facet of economic analysis involved the failure to put a value on lost assets as natural resources were extracted and simply assuming that they were irrelevant to the economic equation. On the whole, economists continued to think of environmental affairs as a drain on the economy rather than a requirement for a healthy economy. This only tended to emphasize the environment as something that had to come later rather than sooner in economic progress.

FAILURES

At the low end of the priority ranking was a set of issues, many of them pertaining to the general problem of growth, on which little progress took place. For the most part these were issues about which the public was little concerned, about which the environmental community had expressed some interest but had not focused on sharply, and about which environmental organizations appeared unable to develop support for action. Others received support in general but with vigorous opposition from those whose personal or corporate behavior would require modification. They reflected a marked failure to make much headway on a set of issues acknowledged to be fundamental environmental concerns.

Basic to these failures is the overall fact that amid the events of the environmental era, population growth and economic development in the United States moved on at a rapid pace and, in their wake, brought a host of accelerated environmental pressures. Environmental successes were confined to smaller and more protective actions rather than extended to larger public objectives. On the whole, the ideology and tenor of the country was to forge full speed ahead in economic growth and to ignore its environmental consequences with the rather rosy general idea that one could have "economic growth and environmental quality" at the same time.

The two broad environmental factors in this equation, population and consumption, seemed to make little dent in the nation's thinking. Population was thought of as a problem in other countries, and environmental organizations confined themselves to supporting funds for family planning abroad, an issue on which they were only partially successful. But they could not bring about a focus on the nation's own population, which continued to grow rapidly in the last half of the twentieth century. Several specialized organizations, such as Environment/Population Balance, Zero Population Growth, and Negative Population Growth, sought to emphasize the issue. But its fragile support was reflected in the difficulty faced by the larger and more diverse environmental organizations in emphasizing it. Especially significant was their failure to bring an environmental perspective to bear on immigration, in which the proposal to reduce the annual number of legal immigrants ran up against a cultural ideology of openness to all peoples that simply ignored the environmental side of the equation. After intense internal debate, for example, the Sierra Club chose simply to say nothing at all about immigration.

A focus on consumption had equally limited success. Appeals to reduce consumption levels fell on deaf ears and became viable only during the high-energy-price years of the 1970s in which there were, for a short while, successful appeals to drive more slowly and consume less gasoline. Handbooks appeared guiding individuals as to how they could consume less, but these had little impact. Consumer recycling carried with it the implication that reuse could make resources go further, but when it was suggested that one way to cope with the waste problem was to produce less waste by consuming less, the strategy had far less popularity than programs to recycle waste once created. The environmental culture had the potential to outline a strategy of frugality, but on the whole, the idea that the nation's consumers should reduce their consumption levels on behalf of environmental objectives had few takers.

Many of these larger issues of population and consumption converged in the topic of urban growth. Cities were vigorous growth machines, promotional centers for more people, a higher level of consumption, and the increased requirements for transportation, communication, housing, and public facilities that growth implied. Driving the impulse were the promoters of development in the private economy and local government leaders for whom development meant higher public revenues that could finance greater public services. But another stimulus was the promoting of more jobs in growing areas, a process that was facilitated in the years after 1970 by the attractiveness of high-quality environments. Studies demonstrated that areas of high environmental quality were areas of high population growth; reports from *Money Magazine,* which rated places to live, reported that air and water quality were top priorities in community preferences. Hence the dynamics of environmental failure involved forces from all sides that undercut almost every effort to restrict urban growth. In only a few places did the public accept the idea of slower growth, a strategy that was often undermined by the claim of adversaries that the advocates really wanted "no growth" rather than "slow growth."

Given the inability to achieve results that coped fundamentally with the amount or pace of growth, most of those concerned with urban growth were confined to "growth management" reforms, where and how growth would take place rather than whether or not it would take place slowly or rapidly. One of the main strategies was to stress the desirability of reducing urban sprawl, that process by which urban population spread out over the countryside around the city, and to direct new development to the already built-up city. This was the strategy of the nation's first significant state land-use law in Oregon, but other states were slow to take this up. Another was to maintain open space in developing areas by cluster zoning, in which housing would be clustered on one small section of a development parcel and the rest would be shared open space for the homes. But perhaps the most permanent "solution," even though partial, was public landownership for environmental objectives or the land conservancy with a similar purpose, both of which took land out of the development market. These strategies, it should be noted, did little to ease the problem of growth and urban sprawl, but only organized it so as to reduce its impact on some areas rather than others.

A variety of environmental issues met with similar failure. There were, for example, many complaints about noise but little noise reduction. Especially striking were failed attempts to restrict transportation noise from airplanes, trucks, and railroads. For highway noise, the response was to erect noise barriers along the highway edges, thus creating corridors for cars and trucks between residential areas. Noise control encountered severe resistance from those who created it, and as a group, they sought to place that subject in the hands of federal authorities with laws that preempted local and state authority. Many a community was frustrated by this simple strategy, which took control of a major element of the lives of its citizens out of its hands. At one time the Environmental Protection Agency had had the authority to balance this loss of control against aid to communities to protect themselves against noise; however, the Reagan administration kept the federal preemptive authority but destroyed the federal assistance.

Equally significant in this rank of effects was the one law that had great promise at the start and that continues on the statute books but that failed miserably, the Toxic Substances Control Act of 1976. This act grew out of the cases of harm to workers from exposure to toxic substances and to consumers who might also be exposed to them. It was formulated amid the general idea that, in the manner in which drugs were treated in the 1938 Food and Drug Act, no toxic substance should be marketed until it had been tested for its safety prior to marketing. The legislation watered this down to the more limited requirement that the Environmental Protection Agency be notified that a particular substance was to be marketed, which might then lead to an EPA requirement for more knowledge about its effects. When passed, it was hoped that the information requirement alone would be a handle on which to exercise some control over human exposure, but although the EPA sought to develop an elaborate administrative system to make it work, the whole effort failed and hence the nation's main law to control exposure to toxic substances fell into relative disuse.

16 ENVIRONMENTAL PERSPECTIVE

The world of environmental politics is filled with competing descriptions and explanations of what it is all about. These competing explanations present the observer of environmental affairs with a dual task: (1) to describe and understand them as integral parts of political debate, each one serving the purposes of one segment or another of that struggle, and (2) to use them as sources to generate a more detached and comprehensive view. The world of environmental debate is so extensive and varied that we cannot round out our treatment of environmental politics without emphasizing the give-and-take of competing facts and ideas as an integral part of the environmental political scene. At the same time, we shall take up the more difficult task of identifying some relatively reliable benchmarks, somewhat detached from the day-to-day combat so as to establish a more adequate perspective on these environmental years.

COMPETING IDEOLOGIES

Environmental affairs have spawned writings galore that try to put a comprehensive spin on the whole, on the order of worldviews. A few of these are written to give substance to the "environmental crisis," that is, the long-term context within which current issues are framed and that defines the actions needed to combat the crisis. Still more are written to give substance to the opposite view, that it is concern with environmental affairs itself that is the crisis and that something must be done to divert that interest in the environment or the society is in peril. Each of these provides a series of "ultimates" to govern thinking. Though these garner considerable attention from those who engage in the give-and-take of ideas, they do not strike home among the great number of people involved in environmental affairs, among whom argument and discussion are more proximate and less ultimate, and issues are cast within a more substantive and less ideological context.

A number of books sustain the idea of an environmental crisis, usually by identifying long historical roots to an evolving and ultimate catastrophe. These were more characteristic of the 1960s and the first part of the 1970s rather than later. At the same time, several more specific aspects of environmental ideologies became popular among the environmental intelligentsia: deep ecology, which argued that the way nature worked independently of humans provided adequate guidelines for human thought and action; ecofeminism, which posited a long-term struggle between the Earth as feminine and human development as masculine and argued that the latter, long dominant, would now give way to the former as environmental affairs triumphed; and the ecological virtue of the Native Americans, which could provide examples of the fundamental virtue of respect for nature that humans now should adopt.

Those who write, read, absorb, and are preoccupied with these ideas are primarily the environmental intelligentsia. Academic institutions contained a wide range of varied environmental specialists, and it was here where most of these formal environmental thinkers were located. They were part of a network of writers, publishers, and book reviewers who sustained mutual ways of thinking. Few scholarly publications arose through which they could express and formulate their ideas, much in contrast to the way in which, in earlier years, socialist and labor movements had spawned a host of opinion journals. One of the few was *Environmental Ethics,* published by a university philosophy department.

The environmental movement as a whole gave rise to a large number of magazines, but these paid limited attention to most of the ideas of the environmental intelligentsia and devoted themselves far more to the ideas and details needed to stimulate action. Environmental actors usually had some notion of a larger crisis, but they were far less inclined to think and ponder over the larger implications and more inclined to engage the world of the environmental opposition. Hence their ideas and actions were shaped primarily by the proximate world of environmental politics.

Anti-environmental ideologies were even more prolific, but their tone and direction was quite different. Some sought to counter the ideology of environmental crisis by affirming that no such crisis existed, but the tone of most of them was to deplore the amount of attention given to environmental affairs and to identify environmental organizations themselves as the main problem. The ideas ranged from arguments that public interest in environmental affairs was aroused and manufactured by the organizations simply to obtain funds for their activities; to arguments that the environmental movement was not a broad public affair but an effort by a small group to impose its will on the rest of society; to arguments that the movement reflected a disturbed mentality.

Some such writers found a place in academic institutions, but the focal point of anti-environmental ideology was a variety of think tanks financed by wealthy benefactors and sometimes related in some way to an academic institution in order to obtain an aura of intellectual respectability. Outlets for journal writings tended to be general anti-government and free-market publications that had a broader subject matter but included environmental subjects. In a few instances, the anti-environmental movement sought to establish a wider magazine audience—with *Our Land,* published in Idaho to focus on land and water resource issues; *Eco,* financed by the corporate regulated industry; and *Garbage,* which in its later issues became transformed into a defense of the regulated industry—but each of these failed to secure a sufficient readership to survive. Hence the anti-environmental think tanks remained the main source of anti-environmental ideological support.

In contrast with the environmental intelligentsia, the anti-environmental intelligentsia found themselves having closer ties with the anti-environmental public, and yet they were also at odds with them. The anti-environmental movement was based on the older extractive industries that were in decline; they tended to cast their resistance to change in ideological rather than pragmatic terms. Hence the ideas of the anti-environmental intellectuals appealed to them. However, most of the anti-environmental movement was rooted in industrial and development enterprises that tended to be enticed by pragmatic rather than ideological politics. Although ideological combat joined together some of the diverse elements of the anti-environmental movement, it separated out the intelligentsia from the pragmatists in much the same way as within the environmental movement itself.

One of the most distinctive thrusts of the anti-environmental intelligentsia was its relationship to environmental science. Scientific knowledge about the environmental world grew steadily; however, the anti-environmental intelligentsia tended to challenge it persistently. Rather than look upon the growth of environmental science in positive and pragmatic terms, they tended to associate it with predictions of tragedy and disaster and instead advanced their own distinctive scientific interpretations. Those ideas tended to be shaped heavily by the political thrust of the anti-environmental movement, which gave rise to that movement's own scientific advisers and writings quite separate from the mainstream of evolving environmental science. Hardly any evaluation of environmental science from this quarter failed to end with an affirmation of the wisdom of the private market, which suggests the close relationship between political ideology and attitudes toward science.

GRASPING THE WORLD OF ENVIRONMENTAL COMPLEXITY

Environmental and anti-environmental ideologies played an important role in environmental politics, but most participants thought about the topic through piecemeal descriptions of events. Different groups of observers described the environmental world through their own circumstances and provided the more independent observer with a number of competing ideas. One can sort out these different "perspectives," identify which part of environmental complexity is brought into focus by which perspective, and identify the distinctive group of observers who sustain each one.

The Nation's Institutional Leaders

The nation's institutional leaders provided an environmental perspective that grudgingly accepted the importance of environmental affairs but amid a belief in the more overriding importance of traditional forms of economic growth and development. They accepted environmental programs as political necessities, required by the rather quirky state of public opinion rather than rooted in firmly established public values and objectives. In their priorities in budgets, programs, public ideas, and speeches they implicitly emphasized the secondary role of environmental affairs. At most they argued that "we can have a good economy and a good environment" at the same time; yet they also spoke of "sustainable development" rather than a "sustainable environment" and rarely identified a satisfactory environment as a firm basis for a satisfactory economy. It was not considered desirable to state such ideas explicitly as priorities, yet their actions daily confirmed them.

Institutional leaders were constrained to argue that though environmental programs in the past had been desirable, they could be improved; yet in making these assertions they generally gave considerable, though guarded, leeway to the affected industries. The most common concession made in this direction came to be the praise of "market forces," the conviction that in some way the market could be made to work on behalf of environmental objectives. This overlay of the "free market" was considered to be ideologically desirable, but when it came to its application, it had more complex twists and turns. One was the notion that one could distinguish between environmental objectives and their implementation; while objectives in the form of standards should be held to firmly, industries could be given greater freedom to implement them. This line between standards and implementation often became blurred in the details of relevant programs such as the so-called XL program fostered by the Clinton administration. In this program, the EPA insisted that greater freedom in implementation be accompanied by steady improvement in the form of progressively lowered emissions. This continual improvement was not what the industries chosen for the program wanted.

A more successful combination of "market forces" and regulation was the "acid rain" provision of the 1990 Clean Air Act. Although that act was touted for its acknowledgment of "market forces" in the form of freely traded emission permits, the most significant element of the act was a legislative requirement that emissions be cut by 50 percent from 1980 levels and be kept there. The trading feature was quite secondary. This "cap and trade" system later came to be applied to nitrogen oxide emissions as the problem of ground-level ozone—of which nitrogen oxides were a major precursor—came to the fore in the mid and late 1990s. In both these cases there was much talk of "market forces," which only obscured the mandatory reductions, but by doing so made them acceptable.

The Environmental Opposition

The environmental opposition avowed a commitment to environmental objectives and yet resisted both extending those objectives and firmly implementing them. They agreed that past environmental legislation was desirable and necessary, but asserted that more extended action was not. Environmental standards were in principle desirable, but their enforcement was excessive and an undesirable burden on private enterprise. Moreover, it was not needed since the environmental attitudes of industry had changed from being hostile to being friendly and hence with this change, enforcement actions should be sharply reduced.

To the environmental opposition, citizen environmental organizations were undesirable and their influence should be curbed. This involved a mixed set of ideas. One was to exaggerate considerably the past influence of such organizations, to argue, as one writer did with considerable approval, that in the 1970s the citizen environmental organizations had marched ahead without serious opposition. Such views were highly exaggerated, since the outcome of environmental debates was more mixed and the environmental opposition had been able to thwart many environmental objectives.

All this constituted a mixed and confused perspective, which often involved a curious juxtaposition of words and behavior. On the one hand, the regulated community expressed a rather fundamental change of "heart" about environmental affairs, and yet, on the other, it continued vigorously to oppose legislation and effective administration much as it had done throughout the last four decades of the twentieth century. The manufacturing industries continued to voice the same opposition to the findings of environmental science as they had since the mid-1960s. The resource-extraction industries continued to oppose with intensity and vigor the allocation of land and water to recreation and ecological purposes. Land developers continued to resist efforts to require them to pay for the social costs of development. Hence the visions of the world that they put forward seemed to say one thing while they put another into practice.

The perspectives of the regulated industries rarely took on the format of rigid ideologies. Arguments that the environmental movement was out to destroy private industry or was an instrument of international conspiracies were heard from this quarter only rarely; nor were arguments heard that the industries were out to destroy the environmental movement. Yet they persistently warned that the next regulatory step, such as the proposed air-quality regulations in 1997, would bring about economic chaos. Such contentions were a product of the intensity of debate and the strategies of public relations specialists concentrating on the momentary issue. The more sharply stated anti-environmental ideologies of the "wise use" movement differed from the more deliberate, no less intense, but less ideological anti-environmental campaigns of the regulated industry.

The Media

The print and electronic media, the most widely followed sources of information and ideas about environmental affairs, provided an environmental perspective heavily shaped by dramatic events and personalities. At the same time, the media provided limited time and resources for reporters to understand the current and historical context of issues, and as a result they were highly dependent on taking quick short cuts to information rather than engaging in careful investigation.

Especially prevailing was the way in which the media reported controversial issues as matters of opinion rather than circumstance. Reporters covered environmental affairs often by obtaining the opinion of one side of a controversy and then of another. Hence the reader's mind was shaped in terms of controversy but with little opportunity to go beyond to the framework of environmental circumstances. Even scientific knowledge the media often reduced to opinion, giving each side, no matter what the more general scientific opinion. The resulting event-opinion format failed to convey one of the most important features of modern environmental affairs, the steady evolution of knowledge about the environment. This practice, in turn, arose from the discovery that "opinion" reporting was the least controversial for the reporter and required the least knowledge about the issue.

Because of their limited knowledge and resources, reporters were vulnerable to the massive public relations campaigns fostered by the environmental opposition. Media reporters provided fairly reliable coverage of the details of readily dramatized events in which the power of the event overshadowed the pronouncements of the public relations specialist. However, when prompted to place events in a setting of larger significance, the context they presented was heavily influenced by public-opinion campaigns from the regulated industry.

Some of this took the form of significant words, such as the report that every new environmental regulation was "strict" when they could well be described as "modest." Some took the form of hostility to particular forms of environmental protest described as NIMBYs, "not in my backyard," when they could just as readily be described in more positive terms as a defense of one's community. In public-lands debates the media tended to emphasize money issues, such as returns to the public treasury from grazing, timber, and mineral activities, and found it more difficult to describe the relevant environmental and ecological circumstances. Or they found it easier to report the sensational details of dramatic tanker oil spills but not the more mundane cases of even larger amounts of oil spilled in less dramatic but persistent ways. They preferred to report deaths rather than illnesses, and hence the entire field of environmentally caused morbidity was less fully treated than were deaths from cancer. And they found it easier to report pollution in terms of effects on human health rather than on nature and wildlife, which they often reported as unreal, antiquarian, or a product of nostalgia and belittled by the phrase "nature lover." Since most environmental issues involved undramatic cumulative change, they were rarely well covered: the effect of fragmentation on habitat, bioaccumulation of toxic chemicals in food, or undramatic cost/benefit issues such as whether the value of life of an older person is less than that of a younger one.

The news media constituted one of the main sources of environmental information, and its distinctive perspective played a major role in shaping the larger public outlook on environmental affairs. It was not simply a matter of the selection of information but the context within which it was cast. The all-pervading characteristic of environmental affairs is the subtle way in which environmental processes work, often embedded in quiet cumulative circumstances, often unobservable to the naked eye. The few overarching "smog" incidents of the years after World War II were well publicized, but most air-pollution sources, transmissions, transformations, and effects were understood only through special measurement and observation. Hence reporting required that the relatively invisible be made dramatically visible.

Scientists

Scientists brought to the environmental scene a rather simple perspective arising from the task of exploring and discovering the environmental world. Here was a vast unknown, the existence of which in earlier years had been sensed only barely. As the exploration began in earnest after World War II, its dimensions began to unfold in rapid order. The environment came to be perceived as all-encompassing, extending into every nook and cranny of the biological world, involving a network of biogeochemical relationships. It also included the world of relationships with the human environment that people established as they grew in numbers, extended their occupancy of the land, air, and water, and increased the intensity of their impact on it. The perspective that all this generated was a continual process of discovery that made the unknown known. The environment generated a never ending opportunity to learn and to nurture a deep sense of aesthetic appreciation for the complex way in which it worked.

This exploration was marked by the discovery of both knowledge and ignorance. One continually made discoveries and at the same time continually faced what one did not know. Ignorance, in fact, expanded faster than did knowledge. The environmental world was so vast, so diverse, so all-encompassing, so interrelated that as soon as one piece of it was pinned down, ignorance about another came into view to require additional exploration.

Controversy over public policy contributed much confusion and distrust within the scientific community and imposed upon scientific inquiry a burden that continually threatened to weaken the scientific impulse itself. The pressure of policy demands on science was often so great that its results were reshaped to fit policy needs or led to the argument that science be cast aside as irrelevant. This divided scientists into two camps: one was committed to continuing the search with a degree of optimism that new discoveries would be of immense value in extending environmental knowledge; the other was cynical about the enterprise, argued that it was given too much weight in public affairs, and joined with other institutions in society to retard it. There was a major paradox here, an affirmation of the need to "have more science" in public decisions, and yet resistance to accept it when discovered. Environmental scientists continually argued for the greater use of environmental science in public decisions, and yet when the critical point came to do so, such as in the application of the Endangered Species Act, a policy in which scientists were highly influential, it was often looked upon as unwarranted.

Environmental exploration by scientists stimulated exploration by many in society at large. Young and old were drawn into the excitement of discovery. This involved no particular planning or organization but arose from a simple connection between inquisitive humans and the enterprise of discovery. Some were prompted to learn more by an experience of environmental crisis that shocked them into action. Often they made their way directly to some source of knowledge, a library with research materials or a scientist in a nearby university or, more recently, resources on the Internet. This induced them to learn more simply as part of their intellectual and emotional development. But no matter its origin, shape, or form, this thirst for knowledge produced a perspective that emphasized the process of making the relatively unknown environmental world more knowable and drew the wider public into the vast enterprise of environmental scientific discovery.

This environmental perspective of scientific exploration came under severe competition from those who were skeptical about the importance of the environmental world. To them, since there were no environmental problems, there was no need to extend or take seriously the discoveries of environmental science. The most important source for such views lay in the political rather than the scientific world, in those quarters who feared the effects of environmental policy on their economic enterprises. Beginning in the 1960s, with each advance in environmental science, the affected business enterprises took a great interest in its discoveries, challenging it at every step, questioning both the methods and research designs that were used and the conclusions that were drawn. These businesses served as a continual retarding force in the advance of environmental science. Moreover, they brought to the table the argument that disputes over the science were contests between qualified and unqualified scientists, or as the argument went, between "good science," which they advanced, and "bad science," which was advanced by the opponents.

The Policy Analysts

Policy analysts concentrated on the task of making judgments about alternative public policies and in so doing contributed their own distinctive perspective to environmental affairs. Their initial step was to focus on the task of evaluating comparatively different public policies to determine their relative rank in public benefit. The techniques for doing this became a profession in itself that was usually thought of in terms of "cost-benefit" analysis or, more recently, the formula of comparative risk. What were the costs and what were the benefits, and did the benefits outweigh the costs? What were the relative risks imposed on humans and the environment by varied environmental circumstances? The methods of analysis imposed on the problem were simple enough but in practice they became more than complex.

Scientific research to determine just what results a policy could bring about was quite limited, and thus in making judgments on the basis of scientific knowledge, one in fact was continually making personal judgments. Benefits, moreover, were matters of public choices about which people continually differed. One could bring some data to bear on the question of whether or not it was better, in terms of cost/benefit ratios, to regulate indoor air pollution or outdoor air pollution, but such a choice was more a matter of disagreement over values that could not be reduced to formulas. Hence though the policy analysts continued to advance their benefit/cost and relative risk formulas as the way to make environmental choices, the perspectives of the experts and the public often went in quite different directions.

Both of these facets of the policy analysts' approach, the adequacy of knowledge to make relative decisions and the choices among objectives, usually became mired in uncertainty and dispute about both questions. The issue then readily came to be: amid such uncertainty, who should make the decisions? The policy analysts were continually tempted to shift such matters to some executive authority independent of the public. This conclusion was based on the generally age-old distrust of the "populace" that governmental and institutional leaders had long maintained. In some respects, therefore, these issues were rather conventional but long-standing issues of power and authority, now transferred to the often nebulous and disputed field of environmental affairs.

The policy analysts had the attention of two influential groups of people. First, the business community applauded their way of looking at the environmental world because they felt that it would modify many an environmental regulation on the grounds that the costs exceeded the benefits. Much of their hope lay not so much in what facts could be brought to bear on the issue but on the possibility of establishing a process by which they could challenge environmental policy. The nation's institutional leaders also provided considerable support to the approach because of their preference for thinking about public policies from a more universal context, which they could associate with the welfare of the nation as a whole. They supported the basic approach of the political economists that the problem was one of choices made among limited resources. Both the business community and the institutional leaders supported the notion that the public's attitudes were "emotional" and hence more than unreliable and that more authority should rest in the hands of centralized decision makers.

Environmental Organizations

The environmental community was the least articulate about the broader contexts of environmental affairs among all the sectors of the environmental scene. Any ideas it generated came more through the implications of action rather than the articulation of formal ideas as a method of argument. Few "think tanks" existed to generate systematic statements of environmental circumstance; instead, there were a great number of different organizations, each representing a distinct type of environmental problem such as air quality, water, wetlands, forests, natural areas, biodiversity, hazardous waste sites, or airplane noise, each giving rise to its own publications and statements of circumstances and issues, but rarely giving rise to what it all meant. Few formal statements of the environmental perspective were forthcoming; rather, those perspectives had to be reconstructed from innumerable actions, sometimes codified in action handbooks.

Their perspective stressed past environmental progress and defined further objectives. Environmental affairs involve continuing and unfinished business, in which general goals have been extensively articulated in the past, and action is devoted to continued and incremental achievement. Definition of ideal and ultimate goals gave way to emphasis on the way in which environmental benefits continue to enhance the quality of life, and these benefits faced continued struggles to make headway amid more dominant developmental drives.

This perspective especially marked the high degree to which environmental engagement is associated with environmental science. This involved a sense of optimism that much can be achieved if one simply works at it, through understanding how the environmental world works and through applying tenacity and knowledge to environmental circumstances. Those active in environmental affairs turn instinctively to science and technology to grapple with environmental circumstances in which they become interested, and those advancing environmental science turn to the environmental public to help in securing the resources both to discover and to implement their findings.

The environmental perspective has markedly failed to articulate fully the central feature of environmental politics, the tension between their own efforts and the environmental opposition. As events succeed events, an awareness of that opposition is ever present and identified in the thick of the battle. But those involved in environmental engagement seem to be most interested in action rather than reflection. They move from event to event with little larger analysis of the present or the past. To those deeply engaged in the search for environmental improvement, the "enemy" is not so much the active opposition but the environmental circumstances that must be challenged and improved. While the "environmental movement" is the subject of many accounts from all sides, few comparable accounts of the environmental opposition have appeared. And the continuing tension between environmental objectives and the environmental opposition remains for the most part obscured.

THE SEARCH FOR A MORE ADEQUATE PERSPECTIVE

Understanding this new phenomenon—environmental affairs—in American society, then, is shaped heavily by the course of the political battles with which it is engaged, and one continually hopes for some perspective that can more adequately inform. Ways of thinking are called for that can place the different views about the environmental world and the human engagement it has engendered in a more overarching context to make sense out of the whole. We have done so in the previous pages, and here we summarize the most important features of such a perspective.

First and foremost is the perspective of "knowledge in the making," of the way in which inquiry into the environmental world has produced a factual basis for thinking that goes beyond the immediate fray of debate and competing visions. Little of that environmental world was known at midcentury, but by the end of the century, enough was known to outline how the environment worked and how humans worked in relationship to it. That knowledge provided the basis for an increasingly firm human perspective. Amid the continuing succession of debated events, the reality of those environmental circumstances could readily be lost to view, but if brought to the fore, it provided relatively solid ground for an overall perspective.

Within this growing knowledge, two benchmarks are especially useful. One establishes a context of multilayered and multifaceted environmental circumstances and the other, an evolutionary context of incremental change over time. As one attempted to reconstruct environmental meaning from the sequence of a multitude of environmental events, it was difficult to go beyond the surface to the context. And yet in almost every circumstance a host of questions that might give contextual meaning to the event were more than obvious if investigators would only probe them. Thus, for example, in the famous alar case in which the Natural Resources Defense Council had issued a report on the effect of pesticide residues in food on child development, the list of peer reviewers in the front of the report could have led reporters, if they were so inclined, to a group of pediatric scientists interested in a range of similar problems, such as the effect of lead exposures, and in turn to a perspective on a long-standing debate within the field of pediatrics as to how to think about childhood diseases. Those who wrote about the report had no such perspective and hence their treatment of the issue was a case of reporting on an event divorced from context.

Each event is embedded in the larger circumstances of human values and their geographical distribution, science, law, relationships among governing institutions, tension between environmental and developmental objectives. A major feature of this book is to bring these contextual factors to the fore to help the reader shape the way in which environmental affairs are perceived. While conventional ways of media reporting prompt one to skip along the visible tips of events, the real world of environmental affairs is an intermixture of deeply intertwined values and institutions. These reinforce each other in such a way that they develop slowly and almost imperceptibly, and, for the most part, without the sensation and drama that the media emphasize. Contextual reconstruction is a fundamental and irreducible requirement for environmental understanding.

Events are embedded not only in context; they are also embedded in the flow of time. Each succeeding event or circumstance is related closely to a preceding one. It is difficult, if not impossible, to understand a point in time without placing it in its evolutionary context. This is not just a view of history that links events past to events present. More important, it is an understanding that emphasizes the way one thing "grows out of" another, it arises from the old and then becomes something a bit different. Media knowledge, with its predisposition to focus on the drama of unusual events, distorts this process by emphasizing how different one event is from another. A bit of scientific knowledge is dramatized by how new it is or by how people once thought one way and now think differently. In the media perspective, incrementalism, slow and steady change, seems to be "off the table" and hence usually falls outside the realm of meaning.

An evolutionary context also suffers from the interests of those who have a stake in the outcome of environmental policy: the nation's institutional leaders, the private industries subject to public regulation, leaders of citizen environmental affairs, and the policy experts whose thought is dominated by their reformist ambitions. In each case, the intense desire to justify policy or to change it leads to an attempt to reconstruct the evolutionary context in order to justify the current proposed action. While the extreme anti-environmentalists posit a conspiracy deeply rooted in the past, the pragmatic environmental opposition of the regulated industries posits an acceptable past in which the role of the environmental public is exaggerated amid an increasingly burdensome present in which the ambitions of the environmental public must be curbed. With all these attempts to shape and mold a historical perspective to fit a current political drive, a more sober and detached account of environmental evolution is needed if an evolutionary perspective more detached from the political fray is to be achieved.

Two strategies used in this book deserve special mention as devices to aid in reconstructing this contextual and evolutionary perspective. One is a comparative method of analysis. Events taken by themselves or only from personal experience can be understood more adequately if taken in comparison of one to another. Thus I have placed the evolution of environmental values and culture in several comparative settings of geography and political party through the analysis of environmental voting patterns in the U.S. House of Representatives. I have sorted out three sets of environmental values—natural values, human health, and ecological stability—

and analyzed each in contrast with the other. I studied the evolution of political tension over environmental affairs through the give-and-take of both environmental impulse and environmental opposition; and I ranked the results of environmental policy through a scale from most effective to least, in which several hundred different policy categories have been arranged by priority. This attempt to describe environmental affairs comparatively and systematically to discover patterns helps to remove the task from the sway both of events and of personal opinion.

An even more essential requirement for a subject of such range and complexity is that observers extend their reach to the far-flung nooks and crannies of environmental affairs and become exposed to evidence that reflects that range. Most observers come to conclusions based on only a limited range of topics among all those involving the natural world, the world of biogeochemical cycles, or the world of economic development. Environmental economics, technology, science, and law must be brought into view as integral parts of the whole. Those who examine policy issues seem to give little attention to the vast number of publications in environmental science, even those that are readily available to the lay person; to environmental policy issues such as are readily available in the many newsletters on a wide range of environmental subjects; or to issues involving the American West, without following evidence about the internal development of the West itself.

The world of contemporary environmental affairs is badly in need of individuals who would reflect on the whole and institutions that could provide the resources for them to do so. Most environmental writing is surrounded either by the task of argument to gain advantage or the task of implementation. Few have time or interest enough to become well informed or to pursue careful understanding of what is one of the most important developments in society and politics in both America and the entire world. Yet just as one can become excited about exploring the world of environmental transformation, one can also become excited about the world of environmental engagement. It is hoped that this book reflects well a lifetime's pursuit of such excitement and can infect others with a desire to continue that pursuit.

17 PAST, PRESENT, AND FUTURE

The ongoing patterns of environmental affairs described in the previous chapters are part of an evolutionary process that links past, present, and future. They emphasize incremental change, in which slowly emerging attitudes and interests, choices in personal values and behavior, public debates and policies add up over the years and take their place amid those inherited from the past. They also establish directions for the future, with a sense of equally incremental change as varied and competing personal and public initiatives work themselves out. Past, present, and future are closely linked and provide a context in which we can understand more fully what is happening around us and help to predict the directions of the future. As a final note, therefore, we place environmental affairs in the broad context of evolutionary change through which we can summarize those links.

ENVIRONMENT AND DEVELOPMENT IN TENSION

Environmental thought and action arose in the midst of a previous and continuing overriding national commitment to development, which heavily shaped environmental affairs. In no sense have environmental concerns become overriding or even reached a place of importance in either personal life or public affairs equal to that of development objectives. Public leaders continue to give major emphasis to the overriding value of population growth, jobs, and investments and measure progress in these traditional terms; only in a few localities and in a few instances have environmental factors made their way to a leading place in the society and politics of the United States. While the nation's institutional leaders give favorable expression to the "environment," they continually warn about too much emphasis on it and implicitly affirm that it is subordinate to conventional economic growth. The dominant view is that it is a luxury, affordable only if sufficient economic growth provides the means to finance it.

The tenuous role of environmental affairs couched among overriding developmental objectives is emphasized through the continual interaction between widespread affirmation of the importance of the environment on the one hand and limited support and often opposition to specific environmental objectives on the other. Institutional leaders on all sides, whether public or private, affirm the importance of environmental objectives. Yet almost immediately after such affirmations they define desirable policy in such a way as to limit its importance. Individuals also affirm the importance of "the environment" but are prone to divorce themselves from its policy implications when they affect them or to be highly selective, affirming them when they fit into their personal agendas but rejecting them when they do not.

It is easy to discuss these paradoxes in terms of inconsistencies in personal thought and values with the objective of discrediting them in argument and debate. But the inconsistencies are more deeply rooted in the complex of values inherent in personal lives and in the society as a whole. Environmental objectives have arisen within the context of overwhelming development objectives and are still subordinated to them. Moreover, it is doubtful that the balance will be reversed and environmental ways of thinking will subordinate developmental interests. Rather, one can predict more accurately that they will remain in tension and conflict for the long run. Environmental objectives will continue to be advocated and to work their way occasionally into the nation's personal and public agendas, but, at the same time, will face the resistance of a dominant developmental agenda.

Amid such continued tension one should not be misled by the continued affirmation of individuals who say "I am an environmentalist" into thinking that this implies that one is fully governed by environmental values. Such self-images do make some sense in terms of degrees of acceptance, but the question is always "in what way" and "how much." One has to be even more skeptical of the continual affirmation from the business community that though it once opposed environmental objectives, it has now "seen the light." Often it is argued that "we are all environmentalists." That phrase usually means that one is willing, even anxious, publicly to affirm the importance of environmental affairs but, at the same time, equally anxious to make clear just which environmental objectives one is willing to support and which ones not.

Predictions for the environmental future, hence, suggest a continuation of the "environmental present" of the last third of the twentieth century, one in which those who press environmental objectives will continue to do so, but also one in which the opposition is ever present, continually honing its political skills to derail as much of the environmental agenda as it can. In no sense can one predict that the "environmental future" would be much different from this continual tension of the last four decades. Over that time developmental and environmental objectives competed vigorously with each other and predictably will continue to do so.

The Nooks and Crannies of Environmental Transformation

The environmental activities of the late twentieth century can be placed against the previous history of environmental transformation in which developmental influences had reigned supreme. Some success has been achieved by those who feel that development needs to be balanced toward environmental objectives. Hence one can summarize much of this environmental activity in terms of the degree to which a new balance has been achieved. The changes are not sweeping and in no measure fully transform the balance but only serve as ways in which environmental initiatives have been able to establish a partially legitimate place within the overriding developmental influences of the nation's personal and public life.

Environmental objectives have advanced not through some massive shift in values but through environmental occupation of the relatively vacant nooks and crannies of the nation's overriding developmental institutions. Nature protection has been selective, shaped heavily by the way in which those active in the nation's economic development in the past found some areas to be unattractive and uneconomic to develop, such as when once having extracted the available natural resources from given areas, they abandoned them; when the works that made development possible became too costly to maintain; or when previous economic activities, such as inland navigation, declined because of competition from new economic ventures. Such cases as these provided the opportunities for environmental uses of the land and water. The main contribution of the environmental era was to produce a group of people interested in environmental uses of resources, so that when opportunities to enhance environmental objectives arose, there were those who were prepared to take them up.

In similar fashion, the new environmental interests were able to become a small part of the production activities of industry. Some industries were willing to absorb the added costs, but in most cases public resources were allocated to finance public benefits. New technologies were advanced by new industries that found profitable markets in providing the means for reducing emissions. The extensive interest of the public in environmental improvement provided some openings as industry sought to improve its public environmental image, but this led only to limited opportunities that were opened further by government actions and demonstrated the restricted opportunities for environmental improvement that came from the private market alone.

One of the most significant openings for environmental objectives was the search for knowledge about the environment and the way it worked. That search, carried out through the newly emerged environmental science, was taken up as a legitimate enterprise by the scientific community and provided an opportunity for the environment to be a subject of continual public interest. As each new study was reported, and although its media significance was usually only momentary and cast in terms of "proof," the cumulative environmental knowledge provided an important context that enabled environmental affairs to occupy a high degree of legitimacy within the nation's public life. Although economic analysts readily subordinated the environment to development as part of their thinking, the scientists did not, and environmental science served as one of the main openings through which environmental affairs continued to be prominent in the nation's thinking.

Some environmental analysts have spoken of major changes in the value systems of modern industrial society in which older developmental values as one "paradigm" are in the process of "giving way" to environmental values as another. This argument has been advanced especially by sociologists, who chart this "paradigm shift" through survey studies. But it is also advanced by institutional leaders who maintain that they themselves have made this shift in values, that they are now "environmentalists" and hence there is no need for further tension and controversy between development and environment. The shift in values lies at the root of much of environmental affairs, but it seems inaccurate to argue that developmental values will "give way" in the future to environmental values or that tension and dispute will fade into the background. Instead, it seems far more accurate to predict that these two sets of values will continue in tandem and in tension for decades to come. When general affirmations of environmental values are turned into specific issues, it becomes obvious that the controversies remain, and there is no reason to believe that those specific issues will decline over the years. The adjustments may differ as time goes on and as environmental values continue to work their way incrementally into the nation's policies, but they will do so only amid the continued vigorous opposition of those whose primary objectives are in economic development.

Persistence and Change in Environmental Affairs

What has changed in the general direction of environmental affairs over the last four decades of the twentieth century and what has not? The relevant environmental conditions gradually evolved from one phase to another in almost dizzying succession. After reaching the larger context of the Earth and its climate, the range of environmental circumstances that were the subject of interest came to some degree of stability. From the early 1980s on, further evolution involved not broad categories of environmental circumstance, but a more detailed and intensive understanding and experience of their characteristics. Hence, as the twentieth century ended, the most significant fact was the greater complexity of the known environment, the greater complexity of actions to cope with environmental circumstances, and the way this complexity shaped the relative political ability and power to influence the outcome of decisions.

In recent decades, public support for environmental objectives remained stable at a relatively high level, with its active initiatives at a somewhat lower level. The environmental opposition remains as it has been since the 1960s, although at a greater level of intensity and determination. Public confidence in citizen environmental organizations as sources of information and action remains high and confidence in the business community low. The pattern of controversies over science, technology, and economics as well as other features of policy remain without significant change. So also do the ever present attempts by management and institutional leaders to find "consensus," ways of resolving disputes based upon the professed belief that the issues are primarily the result of an inability to talk issues out.

Far more significant as a trend is the increasing complexity of the issues and especially the way they are dealt with. The context of politics changed markedly from its simpler format in the 1960s to the far more detailed and complex format of the 1990s. The continued tension and struggle between the two sides of the environmental debate shaped this enhanced complexity of knowledge and action. As the debate proceeds, more knowledge is brought to the fore to provide an ever more elaborate context within which to think about environmental affairs. At the same time, more procedures and ways for the two intractable forces to come to grips with each other are generated. Thus, while the issues and forces remain the same, their battleground has become more complex, forming an even more rocky terrain.

This complexity set the stage for the continuing drama of environmental politics. Political power has now become the ability to wend one's way through these complexities, and the key to that political power is information and the expertise and technologies required to command it. Hence the most intriguing political drama of these years is the continued struggle between the environmental community and the environmental opposition over the control of information. The environmental community continually seeks to make environmental circumstances more visible, while the opposition seeks to make them more obscure. The environmental community continually tries to simplify problems to enable them to be dealt with more effectively, while the opposition tries to make them more complex to retard action. The context of complexity grows and makes its mark primarily on the political inequality between those who advance environmental objectives and those who oppose them.

Human Pressures and Environmental Limits

A final observation to be made is the limited emphasis in environmental affairs on the central environmental circumstance: the increasing human pressure on the finite environment. Economic growth has continued at a rapid pace; population has grown equally rapidly and so has consumption. Almost every facet of development pressures, long in existence, has grown in extent and intensity. In contrast, little headway has been made in identifying, let alone coping with, the fact of limits. Among the proponents of economic growth there is little thought that there are any limits at all; in the area of population growth, though some emphasis has been placed on the need to reduce its growth abroad, there is little thought that this needs to be done in the United States; and though a few environmentalists have sought to focus on the environmental consequences of increasing levels of consumption, the idea has barely dented the consciousness of the American people.

The piecemeal and incremental approach to environmental issues has brought a few cases of limits into public affairs. A recurrent but temporary experience is encountering the limits of open space amid the growth of population and the proliferation of homes and shopping centers. People seek more open space and find it only to experience the press of population that destroys it. A more permanent sense of limits has underscored the drive for permanent protection of natural areas; almost every attempt to establish legally protected wilderness and wild areas, natural rivers, habitat for plant and animal species, and permanent open space has arisen from the belief that unless those areas are insulated from the pressures of the private market, they will be lost forever. The message of such actions is to draw the line and say that these invading and imperialistic development pressures can go only so far.

As anti-pollution programs have evolved incrementally, they have demonstrated that a continually growing amount of pollution cannot be absorbed by a limited amount of air and water. Early air- and water-pollution programs did not assume the fact of limits and hence established emission standards based on pollution concentrations rather than total amounts. With the increase in the number of sources as a result of economic growth, these standards did not reduce the total amounts of emitted chemicals and thus had limited effects on improving environmental quality. Gradually, on a pollutant-by-pollutant basis, action was taken to limit emissions quantitatively: CFCs into a finite ozone layer; sulfur and nitrogen compounds into a finite lower atmosphere; lead in gasoline into a similar finite quantity of air; emissions that reduce visibility into a finite visual field; limitations on hunting and fishing to maintain wildlife populations; limits on the number of people using wilderness areas. Such recognition of finiteness came about not through any general ideas about the finite environment but through dealing with specific issues and specific chemicals.

The historical evolution of environmental affairs has defined more sharply the engagement between environment and development. The tension between them in the foreseeable future will predictably continue much as in the past. The impact of environmental interests on the role of traditional patterns of development within the United States has been modest, selective, and incremental rather than comprehensive. Only in a few and occasional ways have Americans come to terms with the fundamental fact of the environmental world: the finite limits of air, water, and land that define the context within which human life takes place.

FURTHER READING

This guide to further reading about environmental politics consists primarily of books and a few articles that enable the reader to explore selected topics further. The literature is vast, and the few selections in this limited list are suggestive, not exhaustive. More extensive references and explanatory notes can be found in Samuel P. Hays (in collaboration with Barbara D. Hays), *Beauty, Health, and Permanence: Environmental Politics in the United States, 1955–1985* (New York: Cambridge University Press, 1987). The present book as well as the one listed above are based less on secondary works and more on research in primary materials that have been deposited in the Environmental Archives of the Archives of Industrial Society at the University of Pittsburgh and are available to the public.

1. THE SETTING OF ENVIRONMENTAL POLITICS

Donealla H. Meadows, Dennis L. Meadows, Jorgen Randers, and William. H. Behrens III, *The Limits to Growth: A Report for the Club of Rome's Project on the Predicament of Mankind* (New York: Universe, 1972). One of the first, and the classic, statement of the dangers of the increasing human load on the finite environment. The approach is one of policy analysis of what ought to happen in the future rather than an analysis of the past or current circumstances of environmental politics.

2. A HISTORY OF ENVIRONMENTAL TRANSFORMATION

The field of environmental history has given rise to few satisfactory syntheses. The reader will profit more from works about selected topics. The major single source for articles and book reviews in environmental history is the journal *Environmental History,* formerly *Environmental History Review.*

John Garrett Capper and Frank R. Shivers, Jr., *Chesapeake Waters: Pollution, Public Health, and Public Opinion, 1607–1972* (Centreville, Md.: Tidewater Publishers, 1983). Historical background to one of the more significant efforts to improve water quality in a single large body of water.

William Cronon, *Changes in the Land* (New York: Hill and Wang, 1983). A study of the differences between American Indians and European white settlers in their relationship to the environment.

Andrew Hurley, *Environmental Inequalities: Class, Race, and Industrial Pollution in Gary, Indiana, 1945–1980* (Chapel Hill: University of North Carolina Press, 1995). Case study of the practice of disposing of waste in lower-income areas of the city.

Shawn Everett Kantor, *Politics and Property Rights: The Closing of the Open Range in the Postbellum South*. Studies in Law and Economics (Chicago: University of Chicago Press, 1998). A case study of the transition from customary use of the open range to its closure through fencing and use as private property.

Barbara McMartin, *The Great Forest of the Adirondacks* (Utica, N.Y.: North Country Books, 1994). A meticulous history of forest land use in the Adirondacks based on the evidence of land transactions over a long period of time.

Duane Smith, *Mining America: The Industry and the Environment, 1800–1980* (Lawrence: University of Kansas Press, 1987). Evolution of the mining industry, with particular attention to the development of environmental mining policies that gave rise to industry opposition.

James A. Tober, *Who Owns the Wildlife? The Political Economy of Conservation in the Nineteenth Century* (Westport, Conn.: Greenwood Press, 1987). The evolution of nineteenth-century wildlife policy, with particular attention to the question of private versus public ownership of wildlife.

Ann Vilesis, *The Unknown Landscape: A History of America's Wetlands* (Covelo, Calif.: Island Press, 1997). A comprehensive account of the history of wetlands from colonial times to the end of the twentieth century. Emphasizes the transition from the earlier focus on development of wetlands to the later emphasis on their value as natural environments.

Marilyn E. Weigold, *The American Mediterranean: An Environmental, Economic and Social History of Long Island Sound* (Port Washington, N.Y.: Kennikat Press, 1974). The use of Long Island Sound as a case study in environmental history.

Donald Worster, *The Wealth of Nature: Environmental History and the Ecological Imagination* (New York and Oxford: Oxford University Press, 1993). An approach to environmental history that stresses the relationship of humans to nature.

3. THE ENVIRONMENTAL IMPULSE

Inquiry into the sources of interest in the environment has preoccupied many researchers, using a variety of methods ranging from opinion surveys to in-depth interviews to analyses of voting patterns in legislatures.

Richard C. Albert, *Damming the Delaware: The Rise and Fall of Tocks Island Dam* (University Park: Pennsylvania State University Press, 1987). A typical case study of a major issue in the transition of the meaning of rivers from a developmental to an environmental perspective in the twentieth century.

Samuel P. Hays, "Environmental Political Culture and Environmental Political Development: An Analysis of Legislative Voting, 1971–1989," in Hays, *Explorations in Environmental History* (Pittsburgh, Pa.: University of Pittsburgh Press, 1998), 400–417. Analysis of voting patterns in environmental issues in the U.S. Congress from 1971 through 1989; argument that those patterns reflect broad regional and party variations in environmental values and culture.

Ronald Inglehart, *The Silent Revolution: Changing Values and Political Styles among Western Publics* (Princeton, N.J.: Princeton University Press, 1977). A cross-national account of countries in Europe and North America in which environmental values are a significant element in the changing interests of the general public.

Stephen R. Kellert, *The Value of Life: Biological Diversity and Human Society* (Covelo, Calif.: Island Press, 1996). An extensive study of human values about the natural world, with a special emphasis on attitudes toward wildlife. Kellert's work is the most thorough on how humans view wildlife.

Willett Kempton, James S. Boster, and Jennifer A. Hartley, *Environmental Values in American Culture* (Cambridge, Mass.: MIT Press, 1997). A study of environmental values that are broadly shared by the American people; based on in-depth interviewing in the style of anthropological methods.

Lester W. Milbrath, *Environmentalists: Vanguard for a New Society* (Albany: State University of New York Press, 1984). A statement of the rise of a "new environmental paradigm" by one of its most eloquent exponents.

Robert Cameron Mitchell, "From Conservation to Environmental Movement: The Development of the Modern Environmental Lobbies," in Michael J. Lacey, ed., *Government and Environmental Politics* (Washington, D.C.: Wilson Center Press, 1989), pp. 82–113.

Kenneth Sayre, ed., *Values in the Electric Power Industry* (Notre Dame, Ind., and London, U.K.: University of Notre Dame Press, 1977). Value conflicts in proposed energy alternatives.

Laurence H. Tribe, Corrine S. Shelling, and John Voss, *When Values Conflict: Essays on Environmental Analysis, Discourse, and Decision* (Cambridge, Mass.: Ballinger Publishing, 1976). Analysis of a single large environmental issue: the establishment of the Middle Delaware Recreational Area. Emphasizes conflicts in human values as the root of argument.

4. NATURE IN AN URBANIZED SOCIETY

The best method to view the range of interests and activities concerning nature is to examine the shelves of such books in the bookstores, to read nature magazines, or to watch nature programs on television.

Craig Allin, *The Politics of Wilderness Preservation* (Westport, Conn.: Greenwood Press, 1982). A history of the passage of the Wilderness Act of 1964.

Thomas B. Dunlap, *Saving America's Wildlife* (Princeton, N.J.: Princeton University Press, 1988). Evolution of federal policy from annihilation to reintroduction of wildlife.

Jean Craighead George, *There's an Owl in the Shower* (New York: Harper Collins, 1995). One of many, many children's books that emphasize nature appreciation and reflect how deep the value the American people place on nature has become.

Miron Heinselman, *The Boundary Waters Wilderness Ecosystem* (Minneapolis: University of Minnesota Press, 1996). A study of one of the nation's foremost wilderness areas by a scientist long active in the fight to keep it as a wilderness area.

John Naar and Alex J. Naar, *This Land Is Your Land: A Guide to North America's Endangered Ecosystems* (New York: Harper Perennial, 1993). A survey of wildlands throughout North America.

Will Sarvis, "The Mount Rogers National Recreation Area and the Rise of Public Involvement in Forest Service Planning," *Environmental History Review* 18, no. 2 (summer 1994): 41–66. Changes in resource management at Mt. Rogers, typical of many similar situations, were influenced heavily by public involvement.

Louis S. Warren, *The Hunter's Game: Poachers and Conservationists in Twentieth-Century America* (New Haven, Conn.: Yale University Press, 1997). Discusses federal and state controls over taking wildlife.

Laura Waterman and Guy Waterman, *Forest and Crag; A History of Hiking, Trail Blazing, and Adventure in the Northeast Mountains* (Boston: Appalachian Mountain Club, 1989). One of the very few studies of changes in outdoor recreation activities over a long period of time, including both the nineteenth and twentieth centuries. Focus on New England.

Loren Williamson, *Earth Keeping: Christian Stewardship of Natural Resources* (Grand Rapids, Mich.: William B. Eerdmans, 1980). A religious emphasis on the environment that became increasingly articulated toward the end of the twentieth century.

5. THE WEB OF LIFE

The vast literature on sources, pathways, and effects of pollutants generates the sense of environmental networks or that "everything is hitched to everything else."

William Ashworth, *The Late Great Lakes: An Environmental History* (Detroit, Mich.: Wayne State University Press, 1987). A study of the history of the Great Lakes and their current condition.

Brain Balough, *Chain Reaction: Expert Debate and Public Participation in American Commercial Nuclear Power, 1945–1975* (Cambridge and New York: Cambridge University Press, 1991). A study of the relationships between experts and the public in the commercial nuclear power controversy.

Rachel Carson, *Silent Spring* (Boston: Houghton Mifflin, 1962). A classic that was the first popular account of the role of toxic chemicals in the environment.

Theo Colborn, Dianne Dumanoski, and John Peterson Meyers, *Our Stolen Future: Are We Threatening Our Fertility, Intelligence, and Survival?—A Scientific Detective Story* (New York: Dutton, 1996). Emphasizes research findings on both humans and wildlife that add up to wide-ranging adverse effects from toxic chemicals. Could be considered a sequel to Rachel Carson's *Silent Spring* to compare its impact with that of Carson's book.

Barry Commoner, *The Closing Circle* (New York: Knopf, 1971). Commoner was one of the most prolific writers on biogeochemical cycles of contaminants; in his writings he stressed the widespread impact of chemicals entering the environment from industrial processes.

John Cronin and Robert F. Kennedy, Jr., *The Riverkeepers* (New York: Simon and Schuster,

1997). An account of one of the most successful organizations dealing with water pollution. It originated on the Hudson River and now has extended to some twenty "riverkeeper" organizations in different states.

Joyce Eggington, *The Poisoning of Michigan* (New York: Norton, 1980). A major episode of chemical contamination in which polybrominated biphenyls in the form of a fire retardant were mixed into cattle feed in a plant in St. Louis, Mich. It was then eaten by cattle and chickens and spread widely, resulting in the loss of many farm animals and contamination of milk, eggs, breast milk, and other human food.

Environmental Defense Fund and Robert Boyle, *Malignant Neglect* (New York: Knopf, 1979). An argument for the importance of toxic chemicals in the environment.

Samuel Epstein, *The Politics of Cancer* (San Francisco: Sierra Club Books, 1978). A wide-ranging exploration of the controversy over the role of toxic chemicals as a cause of cancer.

Philip L. Fradkin, *An American Nuclear Tragedy* (Tucson: University of Arizona Press, 1989). Controversy over fallout from nuclear tests. A celebrated episode in the impact of nuclear testing on downwind populations in such states as Nevada, Utah, and Arizona.

John W. Gofman and Arthur R. Tamplin, *Poisoned Power: The Case against Nuclear Power Plants* (Emmaus, Pa.: Rodale Press, 1971). Examines the effect of radioactive emissions from nuclear power plants and explosions.

Jon R. Luoma, *Troubled Skies, Troubled Waters* (New York: Viking Press, 1984). One of the many popular accounts of the impact of acid rain.

William Brooks McAfferty, *Air Pollution and Athletic Performance* (Springfield, Ill.: Charles C. Thomas, 1981). One of the few statements of the adverse effects of air pollution on athletes and activities involving the adjustment of athletic practice and training to the circumstances of air pollution.

Andre Mele, *Polluting for Pleasure* (New York: Norton, 1995). Air pollution from the use of motor-powered water craft.

Sharon L. Roan, *Ozone Crisis: The Fifteen-Year Evolution of a Sudden Global Emergency* (New York: John Wiley & Sons, 1989). A history of the gradual impact of scientific research on thinking about the presence of the "ozone hole" in the upper atmosphere.

William E. Sharpe and Joy R. Drohan, eds., *The Effects of Acidic Deposition on Pennsylvania's Forests* (University Park: Pennsylvania State University Press, 1998). Scientific studies of the impact of acid rain on Pennsylvania soils and forests. Includes reports from a team of scientists from Germany and Canada who, prior to a conference at State College, Pa., in September 1999, visited the state's forests and reported on the causes of their decline.

David Zwick, with Marcy Benstock, *Water Wasteland: Ralph Nader's Study Report on Water Pollution* (New York: Grossman, 1971). A book of considerable influence in formation and passage of the Clean Water Act of 1972.

6. LAND DEVELOPMENT

Land-use issues took a wide variety of forms from urban to countryside to wildland areas; almost every environmental issue turned, in some form, on the use of land.

Luther J. Carter, *The Florida Experience: Land and Water Policy in a Growth State* (Washington, D.C.: Resources for the Future, 1972). One of the better books about state conservation programs.

John Clark, *Coastal Ecosystems: Ecological Considerations for Management of the Coastal Zone* (Washington, D.C.: Conservation Foundation, 1974). A major contribution to the issue of land use in the coastal zone.

Beryl Robichaud Collins and Emily W. B. Russell, eds., *Protecting the New Jersey Pinelands: A New Direction in Land-Use Management* (New Brunswick and London: Rutgers University Press, 1988). One of the most important cases of attempts to control development by identifying areas to be developed and those to be left undeveloped. It involves the transfer of development rights as the major method of implementation.

Charles H. W. Foster, *The Cape Cod National Seashore: A Landmark Alliance* (Hanover and London: University Press of New England, 1985). Explores the evolution of one of the most prominent seashores/lakeshores, which were a new form of land protection in the second half of the twentieth century.

David M. Gillilan, *Instream Flow Protection: Seeking a Balance in Western Water Use* (Covelo, Calif.: Island Press, 1997). Outlines the process whereby "instream flow" and its protection for the fishery and recreation enterprises—in contrast to water extraction for commodity purposes—came to be a part of the debate over the use of western rivers.

Robert G. Healy, ed., *Protecting the Golden Shore: Lessons from the California Coastal Commission* (Washington, D.C.: Conservation Foundation, 1978). California displayed the most successful case of coastal preservation, and the California Coastal Commission was its main implementing instrument.

Robert C. Hoffman and Keith Fletcher, *American Rivers: An Assessment of State River Conservation Programs* (Washington, D.C.: River Conservation Fund, 1984). A survey of state river protection programs.

Wallace Kaufman and Orrin Pilkey, *The Beaches Are Moving: The Drowning of America's Shoreline* (New York: Doubleday, 1979). Applies considerable knowledge about moving shorelines to land-use policies much in line with the approach of Ian McHarg.

Jane Holz Kay, *Asphalt Nation: How the Automobile Took Over America and How We Can Take It Back* (Berkeley and Los Angeles: University of California Press, 1997). An enthusiastic analysis of the detrimental impact of the automobile on urban life and an equally enthusiastic statement of how that can be changed.

Marc Karnes Landy, *The Politics of Environmental Reform: Controlling Kentucky Strip Mining* (Washington, D.C.: Resources for the Future, 1976). One of the celebrated cases of the effort both to reclaim already damaged lands and to prevent future damage.

Charles E. Little, *Challenge of the Land: Open Space Preservation at the Local Level* (New

York: Pergamon Press, 1968). One of many books to promote preservation of community open space.

Ian L. McHarg, *Design with Nature* (New York: Doubleday, 1971). A classic on strategies to reduce damage from building on vulnerable locations such as sand dunes and seashores, steep slopes and floodplains simply by not building there or by "designing with nature."

William K. Reilly, ed., *The Use of Land; A Citizen's Policy Guide to Urban Growth* (New York: Crowell, 1973). An early statement of the importance of land-use problems as a focal point for environmental issues. Reilly became head of the Conservation Foundation and, later, administrator of the Environmental Protection Agency during the administration of President George Bush.

7. PUBLIC RESOURCES AND PRIVATE RESOURCES

The debate over private rights and public responsibilities in resource use involves many little recognized facets; one is the long-standing role of public ownership and management.

William Alverson, C. Walter Kuhlmann, and Donald M. Waller, *Wild Forests, Conservation Biology, and Public Policy* (Covelo, Calif.: Island Press, 1994). A study of the impact of ecological thinking on management of the national forests by three individuals involved in attempts to bring such ideas to bear on national-forest management in Wisconsin.

David A. Clary, *Timber and the Forest Service* (Lawrence: University of Kansas Press, 1986). A study by a former official of the U.S. Forest Service that outlines the dominant role of wood production in the agency's history.

Denzel Ferguson and Nancy Ferguson, *Sacred Cows at the Public Trough* (Bend, Ore.: Maverick Publications, 1983). A widely influential statement about the misuse of the public range by the livestock industry.

John C. Freemuth, *Islands under Siege: National Parks and the Politics of External Threats* (Lawrence: University Press of Kansas, 1991). A study of the major concern that evolved in the 1980s about the threat of influences from outside the parks on the parks themselves. This included air pollution, water pollution, and adjacent development that adversely affected the natural qualities of the national parks.

Paul Hirt, *A Conspiracy of Optimism* (Lincoln: University of Nebraska Press, 1994). A history of the wood-production focus in the national forests, with an emphasis on the self-confidence of agency leaders, continually misplaced, that they knew precisely how to manage the forests.

Craig Johnson, Lloyd E. McCaugherty, Jr., and James M. Morrisey, eds., *California's Private Timberlands: Regulation, Taxation, and Preservation* (Stanford, Calif.: Stanford Environmental Law Society, 1973). One of the very few studies of attempts to establish public regulation of private forests.

Robert B. Keiter and Mary S. Brope, eds., *The Greater Yellowstone Ecosystem: Redefining America's Wilderness Heritage* (New Haven, Conn.: Yale University Press, 1991). A study of the impact of ecological thinking on the Yellowstone area; it defined that area as a comprehensive ecosystem extending far beyond the formal boundaries of the park into the adjacent national forests and private lands.

Christopher McGory Klyza, *Who Controls the Public Lands? Mining, Forestry, and Grazing Policies, 1870–1900* (Chapel Hill: University of North Carolina Press, 1996). A general review of public land policies, with an emphasis on influences impinging on them.

William J. McShea, Brian Underwood, and John H. Rappole, eds., *The Science of Overabundance: Deer Ecology and Population Management* (Washington, D.C.: Smithsonian Institution Press, 1997). A report on one of the more devastating influences on the eastern national forests: excessive deer numbers.

Nathaniel P. Reid and Dennis Drabelle, *The United States Fish and Wildlife Service* (Boulder, Colo.: Westview Press, 1984). A study of the federal agency that manages the national wildlife refuge system and supervises many federal wildlife policies.

Richard West Sellers, *Preserving Nature in the National Parks, A History* (New Haven, Conn.: Yale University Press, 1997). A National Park Service official traces the gradual impact of ecological thinking on an agency that for decades was primarily focused on tourists and tourist accommodation.

8. ENVIRONMENTAL ENGAGEMENT

The forms of human environmental interest; organizations and their activities; human appreciative and intellectual activities.

At several times during the environmental era, compilations of environmental objectives were issued. They provide a good sense of the range of those objectives and the perspective behind them. Two of these are John H. Adams, et al., *An Environmental Agenda for the Future* (Covelo, Calif.: Island Press, 1985), and Gerald O. Barney, ed., *The Unfinished Agenda: The Citizens' Policy Guide to Environmental Issues* (New York: Crowell, 1977).

Thomas B. Allen, *Guardian of the Wild: The Story of the National Wildlife Federation* (Bloomington: Indiana University Press, 1987). History of the National Wildlife Federation.

John J. Berger, *Restoring the Earth: How Americans Are Working to Renew Our Damaged Environment* (New York: Knopf, 1985). An account of some of the many restoration projects under way by citizens throughout the country.

Lytton K. Caldwell, Lyton R. Hayes, and Isabel MacWhiter, *Citizens and Environment: Case Studies in Popular Action* (Bloomington: Indiana University Press, 1976). A good account of types of citizen environmental activity in the United States.

Eric A. Goldstein and Mark A. Izeman, *The New York Environmental Book* (Covelo, Calif.: Island Press, 1900). One of the few comprehensive descriptions of environmental conditions of a single city.

Kentuckians for the Commonwealth, *Making History: The First Ten Years of KFTC* (Prestonburg, Ky.: Kentuckians for the Commonwealth, 1991). History of a state environmental group that was able to bring about major changes in the legal context of strip mining.

Stewart L. Udall, *The Quiet Crisis* (New York: Avon Books, 1963); revised as *The Quiet Crisis and the Next Generation* (Salt Lake City: Peregrine Smith Books, 1988). A classic and one of the first calls for environmental policies from a national leader, the secretary of the interior during the administration of Lyndon B. Johnson.

9. THE ENVIRONMENTAL OPPOSITION

Over the years, a considerable number of books have been written to attack the environmental movement. This literature reflects an organized movement as significant as the organized environmental movement itself. However, the subject has attracted few analysts of anti-environmental activity as a whole.

Ron Arnold, *At the Eye of the Storm: James Watt and the Environmentalists* (Chicago: Regnery Gateway, 1982). A sympathetic account of the secretary of the interior during the Reagan administration, who was a key leader in the anti-environmental public lands movement. Written by one of the prominent leaders of the wise-use movement.

Sharon Beder, *Global Spin: The Corporate Assault on Environmentalism* (Dartington, U.K.: Green Books, 1997). One of the first comprehensive analyses of anti-environmentalism, with a broad international focus. Beder is a professional engineer and Senior Lecturer in Science and Technology Studies at the University of Wollongong, Australia, and a writer on environmental affairs well known in Australia, but much less so in America.

Alston Chase, *In a Dark Wood: The Fight over Forests and the Rising Tyranny of Ecology* (Boston: Houghton Mifflin, 1995). A statement by a vigorous opponent of environmental public land policies and their advocates. Chase identifies the problem as a vast conspiracy by ecologically minded people to undermine traditional wisdom. An attack on both ecological science and ecological values.

Gregg Easterbrook, *A Moment on the Earth: The Coming Age of Environmental Optimism* (New York: Viking, 1995). A comprehensive attack on organized environmental activity; argues that most environmental problems have been solved and that in the future no government activity on behalf of environmental objectives will be needed.

John Echeverria and Raymond Booth Ely, *Let the People Judge: Wise Use and the Private Property Rights Movement* (Covelo, Calif.: Island Press, 1995). An analysis of the "private property" movement from the viewpoint of environmental objectives.

Alan M. Gottlieb, *The Wise-Use Agenda: The Citizen's Policy Guide to Environmental Resource Issues* (Bellvue, Wash.: Free Enterprise Press, 1989). A Task Force Report to the Bush administration by the wise-use movement. Gottlieb is one of the most prolific anti-environmental writers; in later books, he focused on the foundations as the culprits because they financed environmental activities. He made a special attack on the Nature Conservancy.

William Tucker, *Progress and Privilege: America in the Age of Environmentalism* (New York: Doubleday, 1982). One of the first major attacks on the environmental movement, using the argument that it is an upper-class movement concerned with its own personal well-being and without concern for the rest of society.

Richard Vietor, *Environmental Politics and the Coal Coalition* (College Station: Texas A&M University Press, 1980). Deals with a range of anti-environmental activities by the coal industry, but it is especially valuable in detailing opposition to surface-mining regulation in Pennsylvania as revealed by an examination of surface-mining permits.

10. THE POLITICS OF ENVIRONMENTAL IMPLEMENTATION

A vast literature about environmental implementation has appeared. Most consists of policy evaluation in which agency performance is examined, followed by the favorite "solutions" proposed by the authors. Little of this literature examines the political context of administrative decisions; much follows the superficial meaning of law and administration and fails to focus on crucial, often little known, decisions of administrative choice.

Douglas J. Amy, *The Politics of Environmental Mediation* (New York: Columbia University Press, 1987). A critical examination of one of the attempts by environmental policy specialists to iron out environmental disagreements through rather traditional mediation devices. Makes clear that the main element in successful mediation is the issue of the relative balance of power that the parties bring to the proceedings.

George Hoberg, *Pluralism by Design: Environmental Policy and the American Regulatory State* (Westport, Conn.: Praeger, 1992). The shift from economic regulation to broader-based social regulation. The author calls the latter "pluralism" because it reflects more diffused political power.

Alan H. Magazine, *Environmental Management in Local Government: A Study of Local Response to Federal Mandates* (New York: Praeger, 1977). One facet of the intricate relationship among federal, state, and local governments in implementing environmental objectives.

Joel A. Mintz, *Enforcement at the EPA: High Stakes and Hard Choices* (Austin: University of Texas Press, 1995). Emphasizes the political choices involved in enforcement. Tends to rely more on public administrative documents and interviews with agency staff rather than the detailed examination of the "paper trail" records in permitting and enforcement.

Barry G. Rabe, *Fragmentation and Integration in State Environmental Management.* (Washington, D.C.: Conservation Foundation, c. 1986). Looks at a crucial aspect of effectiveness in state environmental administration—the tendency of states to fragment administration and thus to undermine an integrated approach to what is an integrated environmental system.

Clifford S. Russell, Winston Harrington, and William J. Vaughan, *Enforcing Pollution Control Laws* (Washington, D.C.: Resources for the Future, 1986). Analyzes enforcement methods in pollution regulation and their effectiveness.

Ben Whitfield Twight, *Organizational Values and the Olympic National Park* (State College: Pennsylvania State University Press, 1983). A study of the role of the U.S. Forest Service in opposing, eventually unsuccessfully, the establishment of Olympic National Park.

John Wargo, *Our Children's Legacy: How Science and Law Fail to Protect Us from Pesticides* (New Haven, Conn.: Yale University Press, 1996). Emphasizes the politics of administering a law; should be read in connection with the Dunlap book on the politics of DDT.

Todd Wilkinson, *Science under Siege: The Politicians' War on Nature and Truth* (Boulder, Colo.: Johnson Books, 1998). Case studies of scientists in environmental administration who sought to bring science to bear on environmental decisions and were then chastised by their superiors for interfering with administrative policy. Such cases at times led to punishment and removal from employment.

11. ENVIRONMENTAL SCIENCE

The role of science, its assessment, and controversies over what is to be studied, how it is to be studied, and the assessment of the results. Few attempts have been made to look at this process comprehensively. The following are efforts "around the edges."

Mary E. Ames, *Outcomes Uncertain: Science in the Political Process* (Washington, D.C.: Communications Press, 1978). Case studies of environmental science disputes as an integral part of environmental politics.

Nicholas A. Ashford and Claudia S. Miller, *Chemical Uncertainty: A Report to the New Jersey Department of Public Health* (photocopy, December 1989). A study of the controversial phenomenon of chemical sensitivity; a typical case of scientists seeking to explore little-known and difficult-to-examine health problems.

E. L. Davis, H. Babich, R. Adler, and S. Danwoody, *Basic Science Forcing Laws and Regulatory Case Studies: Kepone, DBCP, Halothane, Hexane, and Carbaryl* (Washington, D.C.: Environmental Law Institute). Examination of how policies often precede scientific data rather than come after it, and the way in which this is a major factor in advancing scientific work.

Thomas Dunlap, *DDT, Scientists, Citizens, and Public Policy* (Princeton, N.J.: Princeton University Press, 1980). A historical study of the DDT controversy that goes much beyond the limited focus on Rachel Carson's *Silent Spring* to wider facets of the controversy. Does not extend to the continuing controversies in the administration of law and regulation.

Paul R. Ehrlich and Anne H. Ehrlich, *Betrayal of Science and Reason: How Anti-Environmental Rhetoric Threatens Our Future* (Covelo, Calif.: Island Press, 1996). An analysis of the attacks on environmental science in the name of scientific ideas that advocate the free market.

Michael Fumento, *Science under Siege: Balancing Technology and the Environment* (New York: William Morrow, 1993). An attack on environmental science by a vigorous opponent of the environmental movement who argues that environmental science is simply "bad science" and unreliable.

Elizabeth Y. Hayworth and John W. G. Lund, *Lake Sediments and Environmental History* (Minneapolis: University of Minnesota Press, 1984). A report on one of the major ways in which environmental scientists identify and explore evidence of past industrial activities. Such activities often left some residue that still remains and can be examined to describe past environmental changes.

Allan Mazur, *The Dynamics of Technical Controversy* (Washington, D.C.: Communications Press, 1981). A general analysis of the course of controversy over science and technology by an analyst who attempts to find patterns in it.

James C. Peterson, ed., *Citizen Participation in Science and Policy* (Amherst: University of Massachusetts Press, 1984). Examples of the way in which citizens have participated in environmental science activities; an antidote to the customary argument that citizens should confine themselves to policy choices and that investigation of details should be left to the experts.

Kathryn Philipps, *Tracing the Vanishing Frogs: An Ecological Mystery* (New York: St. Martin's Press, 1994). A study that follows scientific inquiry into the decline in frog populations. An excellent case study on how environmental scientific inquiry proceeds.

Joel Primack and Frank von Hippel, *Advice and Dissent: Scientists in the Political Arena* (New York: Basic Books, 1974). A very general statement of the role of scientists in policy choices.

Edmund P. Russell III, "Lost among the Parts per Billion; Ecological Protection at the United States Environmental Protection Agency, 1970–1993," *Environmental History* 2, no. 1 (January 1997): 29–51. The struggle within the Environmental Protection Agency to bring the ecological effects of pollution into agency policy in the face of the agency's dominant focus on human health protection.

12. THE ENVIRONMENTAL ECONOMY

Environmental interests have influenced more traditional economic activities in many and diverse ways; taken together they constitute a distinctive "environmental economy."

Robert D. Bullard, *Dumping in Dixie: Race, Class, and Environmental Quality* (Boulder, Colo.: Westview Press, 1990). Bullard, a writer on the issue of "environmental justice," emphasizes the practice of locating polluting industries, including waste sites, in communities of blacks and Hispanics. This issue dominated much of the popular field of environmental justice toward the end of the twentieth century.

Herman E. Daly, *Steady State Economics: The Economics of Biophysical Equilibrium* (San Francisco: W. H. Freeman, 1977). A broad, general philosophy by an economist who argues that economics as a profession is quite old-fashioned and needs refurbishing around such matters as environmental costs and benefits.

Alan Thein Durning, *Green-Collar Jobs: Working in the New Northwest* (Seattle, Wash.: Northwest Environmental Watch, 1999). Analysis of the "new economy" of the Pacific Northwest, which contains a vigorous environmental component. Valuable as a source of methods of economic analysis as well as conclusions about the regional economy.

John Elkington, Julia Hailes, and Joel Makoner, *The Green Consumer: You Can Buy Products That Don't Cost the Earth*. A statement of the economic benefits of "green consumption" both to the consumer and to the producer.

A. Myrick Freeman, Robert H. Haveman, and Allin Y. Kneese, *The Economics of Environmental Policy* (New York: Wiley, 1973). One of the many statements by "classical economists" as to the way in which economists could modify their approach to take environmental affairs into account. They propose especially to incorporate "external costs," the cost of pollution imposed on the community and the wider society, into its balance sheets.

Paul Hawken, *The Ecology of Commerce: A Declaration of Sustainability* (New York: Harper Collins, 1993). An argument for the ecological basis of a sound economy.

Robert H. Keller and Michael F. Turek, *American Indians and National Parks* (Tucson: University of Arizona Press, 1998). Emphasizes the reduction of jurisdiction by the Indians on their own lands when federal public jurisdiction was established through the formation of the national parks.

Thomas Michael Power, *Lost Landscapes and Failed Economies: The Search for a Value of Place* (Covelo, Calif.: Island Press, 1996). A study of the economy of the North and Northwest that emphasizes the transition from an extractive economy—mining, livestock grazing, wood production—to one oriented more toward services, recreation, and retirement.

Michael G. Royston, *Pollution Prevention Pays* (Oxford: Pergamon Press, 1979). A statement by one of the leaders in promoting efforts by industrial managers to enhance their earnings by reducing costs of pollution cleanup and providing new products. Royston was far more popular in Europe than in America, where his ideas were taken up by only a few individuals and some states under the watchword, "Pollution Prevention Pays."

Peter Self, *Econocrats and the Policy Process: The Politics and Philosophy of Cost-Benefit Analysis* (Boulder, Colo.: Westview Press, 1975). A statement as to the potential value of a broadly based cost-benefit analysis.

Michael Silverstein, *The Environmental Factor: Its Impact on the Future of the World Economy and Your Investments* (New York: Longman, 1990). A statement by an economist who is optimistic about the way environmental affairs stimulates new and beneficial economic activity.

13. ENVIRONMENTAL TECHNOLOGY

A considerable amount of environmental technology to enhance "green production" was developed in the last third of the twentieth century. There is little treatment of this as a whole, and most books and articles are primarily advocacy arguments for a particular "green" technology. Development of air-pollution-control equipment, for example, is one of the major cases of a successful "green" technology, but it is yet to be chronicled.

John S. Allen, *The Complete Book of Bicycle Commuting* (Emmaus, Pa.: Rodale Press, 1981). A comprehensive statement about one of the alternative modes of transportation to the automobile.

Mary Ann Baviello, et al., *The Scrubber Strategy: The How and Why of Flue Gas Desulfurization* (Cambridge, Mass.: Ballinger Publishing, 1982). A book produced amid the intense debate over the wisdom of the new technology of "scrubbers" for sulfur dioxide emissions from utility plants. Perfection of the technology over the years led them to be almost "conventional" by the 1990s.

Daniel Behrman, *Solar Energy: The Awakening Science* (Boston: Little Brown, 1976). A general statement of the potential and predictable future role of solar energy in the society's energy supply.

David Dickson, *The Politics of Alternative Technology* (New York: Universe Books, 1974). The philosophy and politics of technologies that are alternatives to the existing dominant technologies.

Robert Engler, *The Brotherhood of Oil: Energy Policy and the Public Interest* (Chicago: University of Chicago Press, 1977). The role of the major oil companies in influencing national oil and energy policies.

Frank P. Grad, et al., *The Automobile and the Regulation of Its Impact on the Environment* (Norman: University of Oklahoma Press, 1975). A comprehensive view of the environmental impact of the automobile.

Donald Husinger and Vicki Bailey, eds., *Making Pollution Prevention Pay* (New York: Pergamon Press, 1982). Outlines pollution-prevention technologies that are practical.

Amory Lovins, *Soft Energy Paths: Toward a Durable Peace* (San Francisco: Friends of the Earth International, 1977). The classic statement by the nation's foremost advocate of alternative energy supplies.

Paul Maycock and Edward N. Steinwalt, *Photovoltaics: Sunlight to Electricity in One Step* (Andover, Mass.: Brickhouse Publishing, 1981). One of a considerable number of books advocating solar energy and particularly photovoltaics.

14. THE STRUCTURE OF ENVIRONMENTAL POLITICS

Environmental affairs, in incremental ways, have played their part in helping to shape the structure of American political institutions. Each of the following books captures some of this influence.

William Ashworth, *Under the Influence: Congress, Lobbies, and the American Pork-Barrel System* (New York: Dutton, 1981). A popular and energetic account of the role of economic lobbies that advance the interests of their particular firms and segments of the economy in the political debate.

Christopher J. Bosso, *Pesticides and Politics: The Life Cycle of a Public Issue* (Pittsburgh, Pa.: University of Pittsburgh Press, 1987). Changes in pesticide policy since the 1940s, within the context of political controversy; emphasizes both legislative and administrative decisions.

Lydia Dotto and Harold Schiff, *The Ozone War: The Controversy among Scientists, Governments, and Big Business over Our Precious Ozone Layer* (New York: Doubleday, 1978). One of a number of accounts of the intense political and scientific debate over the "ozone hole."

Gordon K. Durnil, *The Making of a Conservative Environmentalist* (Bloomington: Indiana University Press, 1995). A personal statement by a Republican Party leader in the state of Indiana whose views about toxic environmental pollutants changed considerably when he became a member of the International Joint Commission supervising water quality in the Great Lakes.

Thomas More Hoban and Richard Oliver Brooks, *Green Justice: The Environment and the Courts* (Boulder, Colo.: Westview Press, 1987). A review of the role of the courts in environmental affairs.

Jonathan Lash, Katherine Fillman, and David Sheridan, *A Season of Spoils: The Story of the Reagan Administration's Attack on the Environment* (New York: Pantheon, 1984).

James Rathlesberger, ed., *Nixon and the Environment: The Politics of Devastation* (New York: Village Voice, 1972). One of the first evaluations of environmental policy for a given presidential administration; highly critical.

The Sierra Club, et al., *A Conservationist's Guide to National Forest Planning*. A guide for environmentalists to the implementation of the National Forest Management Act of 1976 on behalf of its environmental objectives. Jointly authored and published by the Sierra Club, the Wilderness Society, the National Audubon Society, and the Natural Resources Defense Council. One of a number of such handbooks produced by citizen environmental organizations to foster citizen involvement in implementing environmental policy.

Allan R. Talbot, *Power along the Hudson: The Storm King Case and the Birth of Environmentalism* (New York: Dutton, 1972). The classic case in which environmentalists were given "standing" by the courts to defend their "environmental rights."

John D. Whitaker, *Striking a Balance: Environment and Natural Resource Policy in the Nixon-Ford Years* (Washington, D.C.: The American Enterprise Institute, 1976). The leading architect of environmental policy during the Nixon administration describes it as a "middle-ground" policy.

15. THE RESULTS OF POLICY

A number of short-term assessments of environmental policy have been issued over the years, but few have considered results over the long-run, such as for the last third of the twentieth century. This chapter of the present book is the only known format developed as a strategy for providing such an assessment.

Robert T. Adler, Jessica Landman, and Diane M. Cameron, *The Clean Water Act Twenty Years Later* (Covelo, Calif.: Island Press, 1993). One of the few books that actually assesses the results, both positive and negative, of a particular program over several years.

Charles F. Floyd and Peter J. Shedd, *Highway Beautification: The Environmental Movement's Greatest Failure* (Boulder, Colo.: Westview Press, 1979). The highway beautification movement, popularly associated with Lady Bird Johnson, wife of President Lyndon B. Johnson, was one of the more hopeful environmental activities of the 1960s. Its impetus was successfully negated by the billboard industry.

Frank Lindsay Grant, *Elephants in the Volkswagen: Raising the Tough Questions about Our Overcrowded Country* (New York: Freeman, 1992). Examination of the wide-ranging environmental impact of a rising level of population in the United States.

The President's Committee on Population and the American Future, *Population and the American Future* (New York: New American Library, 1972). Study by a committee on population and the environment appointed by President Richard Nixon, who then rejected the study after it was completed. In succeeding years, the interest in population issues on the part of both the public and policymakers waned. This study remains one of the best on the subject.

16. ENVIRONMENTAL PERSPECTIVE

Although media sources for information about the environment abound, little systematic study has been made of their relative usefulness.

Michael Frome, *Green Ink: An Introduction to Environmental Journalism* (Salt Lake City: University of Utah Press, 1998). Frome, a prolific writer, examines many facets of environmental journalism.

Craig L. LeMay and Everett E. Dennis, eds., *Media and the Environment* (Covelo, Calif.: Island Press, 1991). Collection of articles from varied sources.

Lou Prato, *Covering the Environmental Beat: An Overview for Radio and TV Journalists* (Washington, D.C.: The Media Institute, 1991). An introduction to the problem of environmental reporting on radio and television; a publication by the Environmental Reporting Forum of Washington, D.C.

17. PAST, PRESENT, AND FUTURE

Several authors probe the implications of the basic context of environmental limits.

Fred Hirsch, *Social Limits to Growth* (Cambridge, Mass.: Harvard University Press, 1976). One of the few statements to emphasize social—in contrast with economic or physical—resource limits.

Joel E. Cohen, *How Many People Can the Earth Support?* (New York: Norton, 1995). A far-reaching analysis of the environmental limitations to world population growth.

INDEX